Sebastian Schröter

Complex dynamics in planar two-electron quantum dots

Sebastian Schröter

Complex dynamics in planar two-electron quantum dots

Südwestdeutscher Verlag für Hochschulschriften

Impressum / Imprint
Bibliografische Information der Deutschen Nationalbibliothek: Die Deutsche Nationalbibliothek verzeichnet diese Publikation in der Deutschen Nationalbibliografie; detaillierte bibliografische Daten sind im Internet über http://dnb.d-nb.de abrufbar.
Alle in diesem Buch genannten Marken und Produktnamen unterliegen warenzeichen-, marken- oder patentrechtlichem Schutz bzw. sind Warenzeichen oder eingetragene Warenzeichen der jeweiligen Inhaber. Die Wiedergabe von Marken, Produktnamen, Gebrauchsnamen, Handelsnamen, Warenbezeichnungen u.s.w. in diesem Werk berechtigt auch ohne besondere Kennzeichnung nicht zu der Annahme, dass solche Namen im Sinne der Warenzeichen- und Markenschutzgesetzgebung als frei zu betrachten wären und daher von jedermann benutzt werden dürften.

Bibliographic information published by the Deutsche Nationalbibliothek: The Deutsche Nationalbibliothek lists this publication in the Deutsche Nationalbibliografie; detailed bibliographic data are available in the Internet at http://dnb.d-nb.de.
Any brand names and product names mentioned in this book are subject to trademark, brand or patent protection and are trademarks or registered trademarks of their respective holders. The use of brand names, product names, common names, trade names, product descriptions etc. even without a particular marking in this works is in no way to be construed to mean that such names may be regarded as unrestricted in respect of trademark and brand protection legislation and could thus be used by anyone.

Coverbild / Cover image: www.ingimage.com

Verlag / Publisher:
Südwestdeutscher Verlag für Hochschulschriften
ist ein Imprint der / is a trademark of
OmniScriptum GmbH & Co. KG
Heinrich-Böcking-Str. 6-8, 66121 Saarbrücken, Deutschland / Germany
Email: info@svh-verlag.de

Herstellung: siehe letzte Seite /
Printed at: see last page
ISBN: 978-3-8381-3898-5

Zugl. / Approved by: München, TU, Diss., 2013

Copyright © 2014 OmniScriptum GmbH & Co. KG
Alle Rechte vorbehalten. / All rights reserved. Saarbrücken 2014

Abstract

Quantum dots play an important role in a wide range of recent experimental and technological developments. In particular they are promising candidates for realisations of quantum bits and further applications in quantum information theory.

The harmonically confined Hooke's atom model is experimentally verified and separates in centre-of-mass and relative coordinates. Findings that are contradictory to this separability call for an extension of the model, in particular changing the confinement potential. In order to study effects of an anharmonic confinement potential on spectral properties of planar two-electron quantum dots a sophisticated numerical approach is developed. Comparison between the Helium atom, Hooke's atom and an anharmonic potential model are undertaken in order to improve the description of quantum dots.

Classical and quantum features of complexity and chaos are investigated and used to characterise the dynamics of the system to be mixed regular-chaotic.

Influence of decoherence can be described by quantum fidelity, which measures the effect of a perturbation on the time evolution. The quantum fidelity of eigenstates of the system depends strongly on the properties of the perturbation. Several methods for solving the time-dependent Schrödinger equation are implemented and a high level of accuracy for long time evolutions is achieved.

The concept of *offset entanglement*, the entanglement of harmonic models in the non-interacting limit, is introduced. This concept explains different questions raised in the literature for harmonic quantum dot models, recently. It shows that only in the groundstate the electrons are not entangled in the fermionic sense.

The applicability, validity, and origin of Hund's first rule in general quantum dot models is further addressed. In fact Hund's first rule is only applicable, and in this case also valid, for one pair of singlet and triplet states in Hooke's atom. For more realistic models of two-electron quantum dots an extension of Hund's first rule for unnatural parity states, the alternating rule, is found to be valid. The origin of the rules is closely related to the angular configurations in centre-of-mass and relative coordinates.

Zusammenfassung

Quantenpunkte spielen eine wichtige Rolle bei einer Großzahl an experimentellen und technologischen Entwicklungen. Insbesondere sind sie ein vielversprechender Kandidat für die Realisierung von Quantenbits und anderen Anwendungen der Quanteninformationstheorie.

Das Hookesche Atom mit harmonischem Fallenpotential ist ein experimentell gut bestätigtes Modell und separiert in Schwerpunkts- und Relativkoordinaten. Experimentelle Ergebnisse die im Widerspruch zu dieser Separation stehen fordern eine Erweiterung des Modells, insbesondere eine Änderung des Fallenpotentials. Ein anspruchsvoller numerischer Zugang wird entwickelt um die Effekte eines anharmonischen Fallenpotentials auf spektrale Eigenschaften von zwei Elektronen Quantenpunkten zu untersuchen. Vergleiche zwischen dem Helium Atom, dem Hookeschen Atom und einem Modell mit anharmonischem Potential werden unternommen um die Beschreibung von Quantenpunkten zu verbessern.

Sowohl die klassischen als auch die Quanten-Merkmale von Komplexität und Chaos werden untersucht. Damit kann die Dynamik des Systems als gemischt regulär chaotisch charakterisiert werden.

Der Einfluss der Dekohärenz kann mit Hilfe der Quantentreue beschrieben werden, die den Effekt einer Störung auf die Zeitentwicklung misst. Die Quantentreue von Eigenzuständen des Systems hängt stark von den Eigenschaften der jeweiligen Störung ab. Verschiedene Methoden zur Lösung der zeitabhängigen Schrödingergleichung werden implementiert und ein hohes Maß an Genauigkeit für sehr lange Zeitentwicklungen wird erreicht.

Das Konzept der *Offset-Verschränkung*, die Verschränkung von harmonischen Modellen im nicht-wechselwirkenden Grenzfall, wird eingeführt. Dieses Konzept ermöglicht es verschiedene Fragestellungen zu klären, welche kürzlich in der Literatur in Bezug auf Quantenpunktmodelle mit harmonischen Fallenpotentialen diskutiert wurden. Es zeigt, dass die beiden Elektronen nur im Grundzustand nicht im fermionischen Sinne verschränkt sind.

Desweiteren wird die Anwendbarkeit, die Gültigkeit und der Ursprung der ersten Hundschen Regel in allgemeinen Quantenpunktmodellen untersucht. In der Tat ist die erste Hundsche Regel im Hookeschen Atom nur für ein Paar von Singulett und Triplett Zuständen anwendbar und in diesem Fall auch gültig. Für realistischere Modelle von zwei Elektronen Quantenpunkten ist eine Erweiterung der ersten Hundschen Regel, die alternierende Regel, gültig. Der Ursprung der Regeln kann in Zusammenhang gebracht werden mit den Winkelkonfigurationen in Schwerpunkts- und Relativkoordinaten.

Prepublications

S. Schröter, P.-A. Hervieux, G. Manfredi, J. Eiglsperger and J. Madroñero. *Exact treatment of planar two-electron quantum dots: Effects of anharmonicity on the complexity.* Phys. Rev. B **87** 155413 (2013).

S. Schröter, H. Friedrich and J. Madroñero. *Considerations on Hund's first rule in a planar two-electron quantum dot.* Phys. Rev. A **87** 042507 (2013).

A. L. Frapiccini, A. Hamido, S. Schröter, D. Pyke, F. Mota-Furtado, P. F. O'Mahony, J. Madroñero, J. Eiglsperger and B. Piraux. *Explicit schemes for time propagating many-body wave functions.* Phys. Rev. A **89** 023418 (2014).

Preface

It has been a great pleasure working on my doctoral thesis at the Physik Department in Garching. During this time I had the opportunity to experience the academic world participating conferences and workshops, teaching undergraduate and graduate students, and collaborating with senior scientists. After all it is the people involved in the process that will be kept in my mind. Consequently, I would like to offer my deep acknowledgements to all those who contributed to my work in the following sentences.

Mein erster Dank geht an Harald Friedrich, dafür, dass er mich in seiner Arbeitsgruppe aufgenommen hat und mir die Möglichkeit gegeben hat mein Promotionsvorhaben unter seiner Aufsicht auszuführen. Ich schätze seine Art über Physik zu denken und habe von seiner Exaktheit und der konsequenten Reduktion aufs Wesentliche immer profitieren können. Ich denke ich kann im Sinne aller Mitglieder seiner Arbeitsgruppe sagen, dass wir das familiäre Klima sehr genossen haben.

Desweiteren bedanke ich mich bei Peter Vogl und Andreas Bausch für Ihre Tätigkeit in der Prüfungskommission meiner Promotion.

Ein besonderer Dank gilt auch meinem Betreuer Javier Madroñero, der mich in die spannende Welt der Wenig-Teilchen-Dynamik eingeführt hat. Seine Unermüdlichkeit und Ausdauer waren immer ein Ansporn für mich. Ihm ist es auch zu verdanken, dass ich so viele Erfahrungen auf Auslandskollaborationen und Konferenzen habe sammeln können.

Bei Martin Fink und Tim-Oliver Müller möchte ich mich für das freundschaftliche Zusammensein während unserer gemeinsamen Zeit als Doktoranden bedanken. Ebenso bedanke ich mich bei den früheren Mitgliedern der Gruppe, welche ich noch aktiv miterlebt habe – Johannes Eiglsperger, Moritz Schönwetter und Patrick Raab. Einen wichtigen Beitrag zur Arbeitsgruppe hat im organisatorischen Bereich auch Laura Darabas geleistet, den ich an dieser Stelle insbesondere würdigen möchte.

Je voudrais remercier en particulier Paul-Antoine Hervieux et Giovanni Manfredi de Strasbourg. Chaque fois je suis été chez vous c'était un grand plaisir pour moi et j' ai bénéficié de votre enthousiasme pour la Physique.

In the same way I would like to thank Bernard Piraux, Ana-Laura Frapiccini, and Aliou Hamido in Louvain-La-Neuve. I have greatly profited from your superior knowledge in solving the time-dependent Schrödinger equation.

I am grateful for the financial support I received from the *COST action CUSPFEL*, the *TUM graduate school*, the *Max-Planck-Institut für Physik komplexer Systeme* in Dresden, and the *Wilhelm und Else Heraeus-Stiftung*. Access to the computing facilities of the *Leibniz Rechenzentrum* and the *Institut de Physique et Chimie des Matériaux de Strasbourg* is gratefully acknowledged.

Für die freundliche Unterstützung beim Druck dieser Arbeit bedanke ich mich ganz herzlich bei Franz. Für Ihre Zeit und Hilfe in Bezug auf diese Arbeit und mein Rigorosum bedanke ich mich von ganzem Herzen bei Kirstin, Marc und Evi.

Ohne meine Eltern, Marion und Siegfried, und ebensowenig ohne meine Geschwister, Isabel und Florian, wäre ich nicht so weit gekommen, dafür gilt Ihnen mein besonderer Dank.

Für Ihre konstante Unterstützung danke ich ganz besonders meinen Schwiegereltern Lena und Roland außerdem auch meiner Schwieger-Großmutter Elvira.

Meiner Tochter Sophie möchte ich danken für die Freude, die sie mir, insbesondere während der Entstehung dieser Arbeit, jeden Tag aufs Neue schenkt.

Schließlich danke ich aus tiefstem Herzen meiner Frau Violetta, die mich immer unterstützt hat und ohne deren Hilfe ich es nie bis hierher geschafft hätte.

Sebastian Schröter

München, den 19. August 2013

Contents

Contents	**7**
Introduction	**10**
1 The harmonic quantum dot model and beyond	**17**
1.1 A model for a two-electron quantum dot	17
1.1.1 Experimental properties	17
1.1.2 The planar model	20
1.2 Generalities	22
1.2.1 Basis representations	22
1.2.2 Symmetry of the non-interacting harmonic model	23
1.3 Classification of Hooke's atom	25
1.3.1 Separation of variables	26
1.3.2 Analytical solutions	26
1.3.3 Semiclassical quantisation	27
1.4 First results for Hooke's atom with anharmonic perturbation	28
1.4.1 Spectrum	28
1.4.2 Oscillator strengths	30
1.4.3 Influence of a magnetic field	31
1.5 Summary	32
2 Numerical treatment	**35**
2.1 The Schrödinger equation in matrix representation	35
2.1.1 From the Hamiltonian to a generalised eigenvalue problem	36
2.1.2 From coordinates to operators	37
2.1.3 Symmetries and the numerical basis	38
2.1.4 Matrix representation	41
2.2 Numerical solution of the generalised eigenvalue problem	43
2.2.1 Diagonalisation and storage	43
2.2.2 Coordinate scaling and convergence	44

		2.2.3	Comparison to literature data	46
	2.3	Observables and details of the system		48
		2.3.1	Expectation values of the potentials	49
		2.3.2	Expectation values and oscillator strengths	49
		2.3.3	Densities of the states	50
		2.3.4	Angular momentum quantum numbers	51
	2.4	Classical model		51
		2.4.1	Description and time evolution	52
		2.4.2	Frequency analysis	53
	2.5	Summary		54

3 Effects of the anharmonicity   55
    3.1   Classical considerations 58
    3.2   Measures of quantum complexity 61
          3.2.1   Spectral measures 61
          3.2.2   Avoided crossings 66
          3.2.3   Eigenvector localisation 72
    3.3   Summary 74

4 Time dependence and quantum fidelity   77
    4.1   Quantum fidelity decay 79
          4.1.1   Quantum fidelity 79
          4.1.2   Quantum fidelity for eigenstates 80
    4.2   Numerical methods for the time evolution 80
          4.2.1   Solving the time dependent Schrödinger equation 81
          4.2.2   Fatunla method 82
          4.2.3   Arnoldi method 85
          4.2.4   Advances for an iterative predictor-corrector scheme ... 88
    4.3   Preliminary results for the quantum fidelity 91
          4.3.1   Quantum fidelity susceptibility for eigenstates 91
    4.4   Summary 94

5 Hund's first multiplicity rule   97
    5.1   Hund's first rule in quantum dots 98
    5.2   Harmonic case 99
          5.2.1   Entanglement and separability 100
          5.2.2   Applicability of Hund's rule 106
          5.2.3   Origin of Hund's rule 107

5.3	Quartic case	110
5.4	Further confinements	114
	5.4.1 Full potential case	114
	5.4.2 Planar billiard	115
5.5	Summary	115

Conclusion and Perspectives 118

Appendix 123

A Addenda to the book 123
A.1 Modified atomic units 123
A.2 Parameters of the predictor-corrector scheme 124
A.3 Spectral properties for the full potential and the planar billiard . 125
A.4 Level structure of the harmonic model 129

List of Tables 133

List of Figures 135

Bibliography 137

Introduction

Introduction

Experimental realisations of electrons in semiconductors confined in all three spatial dimensions – quantum dots – were first reported in the 1980's [1, 2, 3]. The typical dimension of few nanometres, which is below the de Broglie wavelength of the electrons, leads to discrete energy levels and particular optical emission and absorption properties. Quantum dots are most frequently realised as nanoparticles of certain semiconductor crystals. State-of-the-art production methods on the one hand offer the opportunity to shape these properties almost arbitrarily and on the other hand make fabrication of large numbers possible [4]. Therefore, quantum dots are particularly interesting for experimental and technical developments [5, 6, 7, 8, 9, 10].

The dynamics of quantum dots resembles the one of atoms and well-established methods from atomic physics can be used for a theoretical description of quantum dots [11, 12, 13, 14, 15]. Similar topics are addressed and the comparison between these similar systems reveals new features. The properties of atomic shell filling in general follow Hund's empirical rules [16, 17]. Highly symmetric few-electron quantum dots can be characterised according to the "periodic table of artificial atoms" and Hund's rules appear to be equally valid [18, 19, 20, 21]. Comparisons between atomic and quantum dot systems [22, 23, 24] have brought up new aspects in the ongoing discussion about the origin of Hund's rules in atoms [25].

In the last few years a lot of effort has been taken in the field of quantum computing [26]. On the theoretical side quantum information gives a detailed description of the possibilities offered by manipulating quantum bits [27, 28]. On the experimental side promising results for the realisations of quantum bits have been reported and a variety of different approaches are considered [29, 30]. Atomic systems, e.g. Rydberg atoms, as well as quantum dots have been proposed as possible realisations of a quantum bit [31, 32, 33, 34, 35]. The main obstacle on the route to technically realised quantum computing is decoherence. Decoherence describes the quantum to classical transition performed by a system (quantum bit) in contact with the environment that forces the system

Introduction

to lose the specific quantal properties. From a modern point of view decoherence is described by the increasing entanglement of the system states with the environmental states [36, 37]. Entanglement plays a crucial role in quantum information and quantum cryptography and theoretical investigations as well as numerical studies of bipartite systems are addressed [38, 39]. Significant differences between harmonic two-electron quantum dot models [40, 41, 42, 43] and atomic helium [44, 45] have been reported [46, 47].

Another way to investigate the decoherence properties of a system is to measure the quantum fidelity or Loschmidt echo [48, 49, 50, 51]. Quantum fidelity is the overlap of a state evolved in time by a Hamiltonian and backwards by the same Hamiltonian with a small perturbation with the original state. Loschmidt argued that a classical statistical system evolved in time up to a certain instant of time, when all momenta of the particles are exactly reversed, would return to its initial state and violate the second law of thermodynamics. At this point classical chaos enters the game. If the system is chaotic, as expected for a classical interacting many-body system, the slightest error in the reversion process will guide it to end up far from the initial state and no violation occurs. This is a consequence of the sensitivity to initial conditions: any initial phase space distance will grow exponentially in time for chaotic dynamics. However, the quantum time evolution is linear and the distance of two vectors, which is the quantum analog to a classical phase space distance, evolved in time under the same Hamiltonian is constant. The alternative idea to perturb the Hamiltonian describing the dynamics as a quantum version of the property of classical chaos had already been proposed by Peres in the year 1984 [52].

The behaviour of the quantum fidelity decay in time is related to the complexity of the system and is well characterised [53]. An unusual behaviour of the decay has been reported by Manfredi and Hervieux for many-body systems investigated with mean-field approaches [54, 55, 56]. This behaviour is particularly interesting since the fidelity stays close to unity for very long times. In other words these systems are robust to environmental perturbations and possible candidates for realisations of quantum bits. A quantum dot system with chaotic dynamics can serve as a model to confirm the unusual behaviour of the fidelity without mean-field approximation.

The fundamental difference between the quantum and the classical time evolution prevents an analogous definition of chaos. What is quantum chaos, if it cannot be defined just like in the classical sense? Different answers have been given: The Bohigas-Giannoni-Schmit conjecture based on random matrix theory predicts universal properties of quantum spectra for systems with underlying chaotic classical dynamics [57]. Semiclassical quantisation of the classical action could be extended to the chaotic case when the regular tori break up. Gutzwiller's trace formula describes the level density in terms of periodic orbits of the classical motion [58]. Semiclassical concepts are seen to be particularly helpful for understanding deviations from the fully chaotic dynamics observed in mixed regular-chaotic systems [59]. Nevertheless, it is not entirely clear how to characterise mixed regular-chaotic systems, neither with established spectral properties [60, 61, 62, 63, 64, 65] nor with alternative concepts [66]. Why is it important to characterise if a quantum system is chaotic? Effects of chaos are omnipresent in classical physics, e.g. in celestial dynamics [67], meteorology [68] or electronic devices like Josephson junctions [69]. Since classical dynamics is the limit of large actions of the quantum dynamics, classical chaos must have a quantum origin. Additionally, classical chaos has become relevant, because it can explain the classical ergodic hypothesis by Boltzmann and therefore the classical creation of thermalisation. In contrast, the connection between quantum chaos and quantum thermalisation is still an open question, just like creation of quantum thermalisation in general [70]. The eigenstate thermalisation hypothesis is an auspicious candidate to explain quantum thermalisation based on Berry's conjecture about universal properties of chaotic eigenstates [71, 72, 73].

The first system for which the ergodic hypothesis was proven in the classical case is the Sinai billiard [74]. Since then planar one-particle billiards have been intensely investigated, as these systems can be well described classically and in the quantum regime [75, 76, 77]. For the characterisation of the quantum properties semiclassical, analytical, numerical and experimental techniques have been presented [78, 79]. Open and closed billiard systems are frequently used as one-particle quantum dot models, but only few approaches with two electrons exist [80, 81, 82]. Experimental realisations and numerical studies of open billiard systems address the correlation of quantum chaos and transport in quantum dots [83, 84]. The theoretical framework interconnecting open and closed billiard systems is summed up under the term leaky billiards [85].

A detailed analysis of a closed quantum dot system with chaotic properties beyond the billiard model has not been performed, yet. It is the goal of this book to shed some light on the complex dynamics, classical and quantum, of such a system. Our investigations are based on a numerical approach for a model of two electrons confined to a plane with full Coulomb interparticle interaction. The effect of a magnetic field perpendicular to the plane is included in the

Introduction

model. The confining potential can be any combination of harmonic and quartic potential. Therefore, a variety of quantum dot models can be addressed, e.g. the well-established Hooke's atom [86], Hooke's atom with a quartic perturbation [87] and the more academic model of a purely quartic oscillator [88].

Spectral properties, expectation values and probability densities can be calculated with high accuracy and are the basis for the investigation of quantum chaos as well as the validity of Hund's first rule. Our model provides sophisticated methods to propagate the system in time. Thus, we can address the effect of small perturbations in long-time evolutions. First steps of the investigation of the unusual behaviour for the fidelity decay are taken [54, 55, 56]. Our framework is in general capable of modelling an equivalent system with an exact numerical approach. Additionally, we obtain results with the analytical analysis of entanglement for harmonic confinement in the non-interacting limit. Calculations of the classical dynamics offer a foundation to our investigations of the quantum complexity of the system. Finally, a perturbative approach for a circular two-electron billiard system corroborates our results concerning Hund's first rule in quantum dots. The flexibility of our approach allows for a detailed investigation of a variety of aspects of the complex dynamics of two-electron quantum dots.

Structure of the present book

The central part of the present book consists of five chapters, each including a short summary. The essence of these chapters is summarised in a concluding chapter followed by the appendix.

The experimental results we consider relevant for our model of a quantum dot are summarised in Chapter 1. We present our model and show that it is capable of reproducing the key features of particular experimentally realised quantum dots.

A detailed description of our numerical approach is given in Chapter 2. We show the analytical transformation of the Schrödinger equation to a generalised eigenvalue problem in a finite matrix representation. This matrix representation is included in our numerical approach and solutions are obtained with standard tools. We analyse the convergence properties and compare to available results in the literature.

In order to characterise the complexity of the system we perform a detailed analysis of the effect of the anharmonic confining potential in Chapter 3. The classical dynamics is studied with an analysis of the dominant frequencies of the trajectories. On the quantum, side three main properties of chaotic systems are addressed: the nearest-neighbour distribution, the distribution of energy gaps of avoided crossings and a localisation property of eigenstates measured by the information entropy of a basis representation.

In Chapter 4 we consider the time dependence of our system with a focus on quantum fidelity. Three different approaches to calculate the time propagation of our system are tested. For a particular iterative predictor-corrector scheme a method to adapt the time-step is developed. Results on the quantum fidelity decay of eigenstates of the unperturbed system are presented.

We will address the applicability and validity of Hund's first rule in detail in Chapter 5. In the course of this chapter the entanglement properties of the harmonic system in the non-interacting limit is related to the applicability of Hund's first rule. This *offset entanglement* can explain some properties of the fully interacting system. We will emphasise the differences between various quantum dot models and discuss the *alternating rule* as an extension of Hund's first rule.

Introduction

Chapter 1

The harmonic quantum dot model and beyond

1.1 A model for a two-electron quantum dot

Electrons in a semiconductor can be confined in all spatial dimensions. The confinement to a plane is in general realised between two semiconducting layers of, e.g. GaAs and AlGaAs, which results in a so-called quantum well. Further confinement is achieved either with gate electrodes or by geometrically shaping the sample, or both. In a quantum wire the dynamics is reduced to a line, and finally, in a quantum dot electrons are confined in all three spatial dimensions. The density of states undergoes a transition from the well-known $\sqrt{E}$-behaviour for the bulk semiconductor, to discrete energy levels for the quantum dot. We will only consider those zero-dimensional objects, so-called "artificial atoms" from now on.

1.1.1 Experimental properties

The physical model we are about to study is based on experimental findings. Since the first experimental realisations of quantum dots in the 1980's [1, 89, 2, 90] there has been a broad interest in these physical nanosystems. The term *quantum dot* was coined shortly after their discovery by Reed [3]. While the first quantum dots contained a huge number of electrons, already in 1989 a controlled number of 3 to 20 electrons per dot was reported [91]. At the beginning quantum dots were only available in huge arrays with statistical fluctuating properties. It was a great success to address and charge single dots with electronic gates [92]. Nowadays quantum dots can be fabricated by various techniques with a great variety of materials and of optical and charging properties. We will discuss some of these properties in this section and connect them to our model.

Before introducing modified atomic units for our approach it is convenient to use SI units for the correspondence between experiments and our model.

Parabolic confinement and number of electrons

Far infrared transmission spectra are obtained experimentally for few-electron quantum dots and show two specific properties relevant for our model [93]. First: The absorption amplitude has discrete values when varying the depth of the potential via an applied gate voltage. This property allows us to determine the number of electrons N on the dot. Therefore N is a good quantum number. Second: The wavelength of the transmission minimum depends only weakly on the number of electrons. Indeed, this has also been shown comparing quantum dots with larger numbers of electrons, e.g. 25 and 100 electrons [94]. For a purely parabolic confining potential a very general result, Kohn's theorem [95], is valid. Such a confinement allows for separation of centre-of-mass (COM) motion independently of the interaction between the electrons. As a consequence Kohn's theorem states that the oscillator strength of the ground state is independent of the interparticle interaction and the number of electrons. We can assume the confining potential to be harmonic with some deviation,

$$V_{\text{conf}}(\boldsymbol{r}) = \frac{1}{2}m^*(\tilde{\omega}_x^2 x^2 + \tilde{\omega}_y^2 y^2 + \tilde{\omega}_z^2 z^2) + V_{\text{anharm}}(\boldsymbol{r}), \quad (1.1)$$

with the effective electron mass m^*. Therefore, we can write a general Hamiltonian,

$$H = \sum_{j=1}^{N}\left(\frac{\boldsymbol{p}^2}{2m^*} + V_{\text{conf}}(\boldsymbol{r}_j)\right) + \sum_{i<j} V_{\text{int}}(\boldsymbol{r}_i, \boldsymbol{r}_j), \quad (1.2)$$

based on the assumption of a discrete number of electrons on the dot and dominantly harmonic confinement. The influence of the confining solid state system on the Coulomb repulsion between the electrons is manifested through the relative permittivity ϵ_r,

$$V_{\text{int}}(\boldsymbol{r}_i, \boldsymbol{r}_j) = \frac{q_e^2}{4\pi\epsilon_r\epsilon_0|\boldsymbol{r}_i - \boldsymbol{r}_j|}, \quad (1.3)$$

with the electron charge q_e and the vacuum permittivity ϵ_0.

Cylindrical symmetry and effectively planar dynamics

Kouwenhouven and coworkers have realised and intensely studied radially-symmetric, few-electron quantum dots [18, 19, 96, 97, 20]. The necessary energy for adding another electron onto the dot can be measured and gives rise to the "periodic table of artificial atoms" [19]. This shell structure resembles the one for real atoms with one major difference: The magic numbers correspond to a two-dimensional dynamics. The qualitatively different confinements mentioned above render the dynamics of a quantum dot effectively two-dimensional, $\tilde{\omega}_z \gg \tilde{\omega}_{x,y}$. Deviations from this assumption will be addressed in the last paragraph of this Section. In the following all vectors are understood to be two-dimensional, unless stated otherwise. Furthermore the angular momentum is a good quantum number and the well-known rules of Friedrich Hund [16, 17] also apply here [21]. We adapt our model to these realisations and assume the dynamics to be two-dimensional and the confining potential to be radially symmetric,

$$V_{\text{conf}}(\boldsymbol{r}) = V_{\text{conf}}(r) = \frac{1}{2}m^*\tilde{\omega}^2 r^2 + V_{\text{anharm}}(r), \qquad (1.4)$$

with $r = |\boldsymbol{r}| = \sqrt{x^2 + y^2}$.

Influence of a magnetic field

Most of the experimental realisations include the application of a magnetic field perpendicular to the plane of the two-dimensional dynamics. Such a magnetic field, $B_z = B$ has two effects on the general Hamiltonian Equation (1.2), an enhanced harmonic confinement and an energy shift according to the Zeeman effect. The harmonic frequency changes to $\tilde{\omega}^2 = \tilde{\omega}_0^2 + \tilde{\omega}_B^2$, with $\tilde{\omega}_B = \frac{q_e B}{2m^*}$, the Larmor frequency. The Zeeman term in the absence of spin-orbit coupling reads

$$\tilde{\omega}_B(L_z + g^* S_z), \qquad (1.5)$$

with the total angular momentum perpendicular to the plane L_z, the z-component of the total spin S_z and the Landé factor $g^* \approx 2$. The larger length scale of quantum dots compared to atoms makes effects accessible in the laboratory that cannot be accessed for atoms. A well-known example is the transition of the ground state from singlet to triplet spin symmetry depending on the magnetic field. This effect had been proposed theoretically [86] before it was measured for quantum dot helium for magnetic fields of a few Tesla [98, 99, 96]. For atomic helium the necessary magnitudes of the magnetic field are five to six orders larger (10^6 T) and are expected to occur only close to astronomical objects [100].

Chapter 1. The harmonic quantum dot model and beyond

Dimensional phenomena and deviations from the parabolic confinement

We take a closer look at two of the assumptions made above, the dimensionality and the harmonicity of the confinement. An anisotropic three-dimensional harmonic model with $\tilde{\omega}_z > \tilde{\omega}_0$ has been investigated with semiclassical methods [101]. By varying the Larmor frequency a smooth transition from effectively two-dimensional, $\tilde{\omega}_z > \tilde{\omega}$, to three-dimensional dynamics, $\tilde{\omega}_B \approx \tilde{\omega}_z$, can be modelled. Confirming the intuitive picture, the Coulomb interaction is weaker in three dimensions, since the electrons have the opportunity to avoid vertically. As a consequence scaling the interparticle interaction can partly account for the effect of the third dimension in the two-dimensional approach.

Another three-dimensional approach calculated the electrostatic potential induced by a charge-free region for $|z| < z_0$ and uniformly distributed charges beyond, $|z| > z_0$ [21, 102]. The quantum dot was cylindrically symmetric and the Poisson equation can be solved analytically. Expanding the potential for small values of r and in the limit $z = 0$ gives a parabolic confinement with quartic perturbation. The strength of the quartic perturbation in this derivation depends on the distance and can be positive or negative. Nevertheless a quartic term can account for the dimensional effects to a certain extent.

Deviations of the far-infrared spectroscopy properties from the behaviour corresponding to purely harmonic confinement had been reported for few- and many-electron quantum dots [93, 94]. A quartic perturbation has successfully been made use of for partially describing these deviations [12]. Following these findings we establish our anharmonic perturbation to be a quartic potential,

$$V_{\text{anharm}}(r) = \tilde{\kappa} r^4. \tag{1.6}$$

We will show in Section 1.4 how this perturbation does not change the experimental validity of our model in comparison to the harmonic confinement. Nonetheless, we will show in Chapter 3 how the dynamics and the complexity of the system change significantly.

By the choice of atomic units variations of the interparticle interaction strength will be mapped to variations of the confining potential strengths. Therefore effects of the dimensionality will be incorporated by varying κ and ω_0 and we will not explicitly consider effects of the dimensionality.

1.1.2 The planar model

The two-electron dynamics shows a high degree of complexity at a performable amount of numerical effort and describes the perfect balance between feasibility and complexity. As mentioned above, the number of electrons can be controlled experimentally and we will restrict ourselves to the case $N = 2$ in the following. The resulting equations resemble those of a description of atomic helium under the assumption of infinite nucleus mass, but different confining potential. The term quantum dot helium is commonly used and is more general than the term Hooke's atom[1], which only corresponds to purely harmonic confinement.

We describe the complex dynamics of a planar two-electron quantum dot with harmonic and quartic confinement, Coulomb interaction between the particles and a magnetic field perpendicular to the plane. Each electron has an effective mass m^* and the influence of the confining solid state system on the Coulomb repulsion is manifested through the permittivity $\epsilon = \epsilon_0 \epsilon_r$. Affected by these two quantities we introduce modified atomic units (a.u.), see Appendix A.1, and the most general Hamiltonian applied reads as

$$H = \sum_{j=1}^{2} \left[-\frac{1}{2}\nabla_j^2 + \frac{1}{2}\left(\omega_0^2 + \omega_B^2\right) r_j^2 + \kappa\, r_j^4 \right] + \omega_B L_z + \frac{\gamma}{r_{12}} + H_{\text{spin}}, \qquad (1.7)$$

with $r_{12} = |\boldsymbol{r}_1 - \boldsymbol{r}_2|$. Note that the Coulomb strength parameter γ is in general unity in our units (a.u.), but we keep it as a parameter to explicitly investigate the effect of the interparticle interaction.

Wave functions and spin properties

The Hamiltonian allows for separation of spatial and spin degrees of freedom. Therefore the wave function describing our system can be written as the tensor product

$$|\psi\rangle = |\boldsymbol{r}_1, \boldsymbol{r}_2\rangle \otimes |S, S_z\rangle, \qquad (1.8)$$

where $|\boldsymbol{r}_1, \boldsymbol{r}_2\rangle$ depends only on the spatial degrees of freedom. The spins of the two electrons couple to the total spin S with z-component S_z. The total wave function must obey the Pauli principle and thus be antisymmetric under particle exchange. The spin wave function for the singlet state,

$$|0,\ 0\rangle = \frac{1}{\sqrt{2}}(|\uparrow\downarrow\rangle - |\downarrow\uparrow\rangle), \qquad (1.9)$$

---

[1] Notice, that the term Hooke's atom is used for the two- and three-dimensional model, while we will in general refer to the planar case in this thesis.

is antisymmetric under particle exchange. The spatial wave function is therefore symmetric $|\boldsymbol{r}_1, \boldsymbol{r}_2\rangle = |\boldsymbol{r}_2, \boldsymbol{r}_1\rangle$. The spin wave functions of the triplet states,

$$|1, -1\rangle = |\uparrow\uparrow\rangle,$$
$$|1, 0\rangle = \frac{1}{\sqrt{2}}(|\uparrow\downarrow\rangle + |\downarrow\uparrow\rangle),$$
$$|1, 1\rangle = |\downarrow\downarrow\rangle,$$

are symmetric under particle exchange and the spatial wave function must be antisymmetric $|\boldsymbol{r}_1, \boldsymbol{r}_2\rangle = -|\boldsymbol{r}_2, \boldsymbol{r}_1\rangle$.

We will only consider the influence of the spin insofar as we will use properly symmetrised spatial wave functions. In particular, we will not distinguish between the different z-components of the total spin, so we set $S_z \equiv 0$ without loss of generality. As a consequence we will omit the spin-dependent term H_{spin} from now on, which is only non-zero for non-vanishing z-component of the total spin and non-vanishing magnetic field.

1.2 Generalities

We can characterise the state of the system based on properties of the non-interacting harmonic model. We will briefly summarise these properties, which are fundamental for the complete analysis presented in this thesis.

1.2.1 Basis representations

The state of our system will be described by the wave function depending on the spatial degrees of freedom. As long as no Coulomb interaction couples the two electrons we can treat them as independent particles. The confinement for each electron is radially symmetric so we can choose a principal quantum number n_j and an angular momentum quantum number m_j, with $j \in \{1, 2\}$ denoting the electron. Taking into account the symmetry under particle exchange we can write the basis of independent particles,

$$|n_1, m_1, n_2, m_2\rangle_{\text{idp}}^{\epsilon_s} = \frac{1}{\sqrt{2}} \left(|n_1, m_1, n_2, m_2\rangle_{\text{idp}} + \epsilon_s |n_2, m_2, n_1, m_1\rangle_{\text{idp}} \right). \quad (1.10)$$

Spin singlet states correspond to $\epsilon_s = +1$, while spin triplet states correspond to $\epsilon_s = -1$. In order to guarantee uniqueness of the representation we impose restrictions on the quantum numbers in such a way that either $n_1 > n_2$ or else

1.2. Generalities

if $n_1 = n_2$ then $m_1 > m_2$. Naturally for $n_1 = n_2$ and $m_1 = m_2$ there is only a singlet state

$$|n_1, m_1, n_1, m_1\rangle^+_{\text{idp}} = |n_1, m_1, n_1, m_1\rangle_{\text{idp}}. \tag{1.11}$$

The quantum number of the total angular momentum perpendicular to the plane

$$L_z|n_1, m_1, n_2, m_2\rangle^{\epsilon_s}_{\text{idp}} = m|n_1, m_1, n_2, m_2\rangle^{\epsilon_s}_{\text{idp}}, \tag{1.12}$$

is the sum of the independent-particle angular-momenta quantum numbers, $m = m_1 + m_2$.

Alternatively the system also separates in COM and relative motion as long as there is no anharmonic confinement ($\kappa = 0$), see details in Section 1.3.1. Associated to each of the resulting radially symmetric subsystems there is a principal quantum number n_j and an angular momentum quantum number m_j with $j \in \{c, r\}$ for COM and relative motion, respectively. A natural basis is thus

$$|n_c, m_c, n_r, m_r\rangle. \tag{1.13}$$

The spin symmetry can be determined by $\epsilon_s = (-1)^{m_r}$, which becomes clear, when considering that particle exchange is equivalent to a rotation of π in the space of the relative motion. The total angular momentum is the sum of the angular momenta in the subsystems, $m = m_c + m_r$.

Due to the two-dimensional confinement the total angular momentum quantum number can take any integer value $m \in \mathbb{Z}$. The case $m = 0$ is special. For even dimensions, two in our case, the general parity operation of changing the signs of all spatial coordinates does not change the orientation of the space. We will address this subtle question of a properly suited parity operation, in particular for vanishing-angular-momentum states in the following Section.

1.2.2 Symmetry of the non-interacting harmonic model

The harmonic, non-interacting Hamiltonian,

$$H = -\frac{1}{2}\left(\boldsymbol{\nabla}_1^2 + \boldsymbol{\nabla}_2^2\right) + \frac{\omega_0^2}{2}\left(r_1^2 + r_2^2\right), \tag{1.14}$$

can be solved easily in several ways, due to the high degree of symmetry of this system. In particular we can find solutions according to the two bases (1.10) and (1.13). To gain insight into the level structure of the general system (1.2), it is instructive to consider the energy levels of (1.14) and their degeneracies.

Chapter 1. The harmonic quantum dot model and beyond

The energy levels are given by $E_n = \omega(n+2)$ introducing the principal quantum number $n \in \mathbb{N}_0$

$$n = 2(n_1 + n_2) + |m_1| + |m_2| = 2(n_c + n_r) + |m_c| + |m_r|. \tag{1.15}$$

The degeneracy of each of these levels is $\#(E_n) = \frac{1}{6}(n+1)(n+2)(n+3)$ leading to a total number of $\frac{1}{24}(N+1)(N+2)(N+3)(N+4)$ states with principal quantum number $n \leq N$.

We will address two different parity operators, the coordinate interchange Π_{xy} and the reflection about the x-axes Π_x. Both change the orientation of the coordinate space and do not commute with the angular momentum operator L_z and thus we cannot build a common eigenbasis. Our concern is to find another symmetry operator to represent all symmetry classes for vanishing-angular-momentum states. Let us consider the action of the two parity operators on the independent particle basis (1.10),

$$\Pi_{xy}|n_1, m_1, n_2, m_2\rangle_{\text{idp}}^{\epsilon_s} = |n_1, \text{-}m_1, n_2, \text{-}m_2\rangle_{\text{idp}}^{\epsilon_s}, \tag{1.16}$$

$$\Pi_x|n_1, m_1, n_2, m_2\rangle_{\text{idp}}^{\epsilon_s} = \tilde{\epsilon}_p|n_1, m_1, n_2, m_2\rangle_{\text{idp}}^{\epsilon_s}, \tag{1.17}$$

with[2] $\tilde{\epsilon}_p = (-1)^{(n-m)/2} = \pm 1$. The total angular momentum quantum number m is even (odd) if and only if the principal quantum number n is even (odd). One can show, that the Π_x-parity operator can only distinguish between negative and positive angular momenta for odd values of $|m|$, which excludes $m = 0$. In contrast the Π_{xy}-operator can be used to symmetrise the bases,

$$|n_1, m_1, n_2, m_2\rangle_{\text{idp}}^{\epsilon_p, \epsilon_s} = \frac{1}{\sqrt{2}}\left(|n_1, m_1, n_2, m_2\rangle_{\text{idp}}^{\epsilon_s} + \epsilon_p|n_1, \text{-}m_1, n_2, \text{-}m_2\rangle_{\text{idp}}^{\epsilon_s}\right), \tag{1.18}$$

$$|n_c, m_c, n_r, m_r\rangle^{\epsilon_p} = \frac{1}{\sqrt{2}}\left(|n_c, m_c, n_r, m_r\rangle + \epsilon_p|n_c, \text{-}m_c, n_r, \text{-}m_r\rangle\right), \tag{1.19}$$

with $\epsilon_p = \pm 1$. We call states with $\epsilon_p = +1$ even and those with $\epsilon_p = -1$ odd parity states. This symmetrisation is possible for all angular momenta, in particular for $m = 0$. For the non-vanishing angular momenta one can either let $m \in \mathbb{N}$ with parity ϵ_p, or $m \in \mathbb{Z}$ without parity. For considerations including a magnetic field only the latter choice gives an eigenbasis of the system. Tables A.5, A.6 and A.7 in the Appendix A.4 give an overview of the number of states in the different symmetry classes.

---

[2]This can be seen by considering a Cartesian basis representation, where $\tilde{\epsilon}_p = (-1)^{n_{y_1}+n_{y_2}}$ for states $|n_{x_1}, n_{y_1}, n_{x_2}, n_{y_2}\rangle$, defined in the usual manner.

1.3 Classification of Hooke's atom

Turning on the Coulomb interaction ($\gamma = 1$) between the electrons turns the simple harmonic oscillator in four dimensions (1.14) into Hooke's atom. This model possesses some unique properties. The Hamiltonian can still be separated in relative motion and COM motion, where the latter is a purely harmonic oscillator in two dimensions. The relative motion is radially symmetric and one can reduce the Schrödinger equation to a radial equation in one degree of freedom. This radial equation can be solved approximately, e.g. with semiclassical methods, and even allows some exact analytical solutions, in particular when applying a magnetic field.

As already mentioned it is the most widely used model of a two-electron quantum dot. Moreover, its simple structure and the similarity to the dynamics of the helium atom makes it a playground for theoretical investigations and in particular for comparisons to the helium atom [103, 104, 23]. There are basically three main differences between these two systems. No continuum nor resonances appear in Hooke's atom, since it is a closed system, i.e. the asymptotic behaviour is completely different. The solutions of atomic helium exhibit the well-known Kato cusp for atomic systems due to the Coulomb singularities of the confinement and the interaction. The solutions of Hooke's atom are smooth for the COM motion, but the solutions of the relative motion are not, corresponding to the Coulomb interparticle interaction. For radially symmetric harmonic confinement Hooke's atom shows regular dynamics in contrast to atomic helium, where the dynamics is mixed regular-chaotic. Especially the last point is of some concern for our model. The introduction of the quartic potential renders the dynamics to be mixed-regular chaotic. We will focus on these properties of the system in Chapter 3. Another wide range of considerations uses Hooke's atom to test the accuracy of approximate methods to solve the Schrödinger equation [105, 12, 40]. The applicability and the origin of Hund's first rule have been addressed [106, 24]. We will give some comments and extension on this treatment in Section 5.2. Furthermore the entanglement between the two electrons has been investigated [40, 46, 43], an issue we will associate with the applicability of Hund's first rule in Section 5.2.1.

1.3.1 Separation of variables

For the sake of completeness we will briefly consider the COM motion. The Hamiltonian after separation of the COM polar coordinates (R,Φ) reads[3],

$$H_{\text{COM}} = -\frac{1}{2}\left(\frac{1}{R}\partial_R + \partial_R^2 + \frac{\partial_\Phi^2}{R^2} + \frac{1}{2}\left(\omega_0^2 + \omega_B^2\right)R^2\right) - i\omega_B\partial_\Phi. \qquad (1.20)$$

The Fock-Darwin levels are the eigenenergies of the corresponding Schrödinger equation [107],

$$E_c = \sqrt{\omega_0^2 + \omega_B^2}\left(2n_c + |m_c| + 1\right) + \omega_B m_c, \qquad (1.21)$$

associated to the eigenstate $|n_c, m_c\rangle$.

A more difficult problem is posed by the relative dynamics, which we consider in the polar coordinates (ρ, φ),

$$H_{\text{rel}} = 2\left[-\frac{1}{2}\left(\frac{1}{\rho}\partial_\rho + \partial_\rho^2 + \frac{\partial_\varphi^2}{\rho^2}\right) + \frac{1}{2}\frac{\omega_0^2 + \omega_B^2}{4}\rho^2 + \frac{\gamma}{2\rho} - \frac{i}{2}\omega_B\partial_\varphi\right]. \qquad (1.22)$$

The usual ansatz for a radial wave function in two dimensions,

$$\psi(\rho, \varphi) = \frac{e^{im_r\varphi}}{\sqrt{2\pi}}\frac{u(\rho)}{\sqrt{\rho}}, \qquad (1.23)$$

results in the radial Schrödinger equation for the wave-function $u(\rho)$:

$$\left[-\frac{1}{2}\partial_\rho^2 + \frac{m_r^2 - \frac{1}{4}}{2\rho^2} + \frac{1}{2}\frac{\omega_0^2 + \omega_B^2}{4}\rho^2 + \frac{\gamma}{2\rho}\right]u(\rho) = \left[\frac{\epsilon_r}{2} - \frac{1}{2}\omega_B m_r\right]u(\rho). \qquad (1.24)$$

In the following we will have a closer look at the solutions of this equation.

1.3.2 Analytical solutions

Starting from an analysis in three dimensions without magnetic field [108] a discrete number of analytical solutions for Equation (1.24) can be found [109]. A power series

$$u(\rho) = e^{-\frac{1}{2}\omega_r\rho^2}(\sqrt{\omega_r}\rho)^{|m_r|+\frac{1}{2}}\sum_{\nu=0}^{n-1}a_\nu(\sqrt{\omega_r}\rho)^\nu, \qquad (1.25)$$

---
[3]Notice that we have performed a scaling transformation, such that $R = \sqrt{2}\,R_{\text{COM}}$ in order to obtain the harmonic oscillator Hamiltonian in the usual form.

is a solution for particular values of the total harmonic frequency $\omega_r = \frac{1}{2}\sqrt{\omega_0^2 + \omega_B^2}$. For example, the frequencies and the associated eigenenergies for the solutions with two $(n = 2)$ and three $(n = 3)$ terms are given by

$$\omega_r^{n=2} = \frac{1}{2(2|m_r| + 1)}, \qquad \epsilon_r^{n=2} = \frac{|m_r| + 2}{2|m| + 1} + \frac{1}{2}m_r\omega_B. \qquad (1.26)$$

$$\omega_r^{n=3} = \frac{1}{4(4|m_r| + 3)}, \qquad \epsilon_r^{n=3} = \frac{|m_r| + 3}{2(4|m| + 3)} + \frac{1}{2}m_r\omega_B. \qquad (1.27)$$

We use these solutions to test the convergence of our numerical method in Section 2.2.3. The appearance of a countable number of analytical solutions has been explained by a hidden symmetry of the system [110]. Furthermore, also closed forms of general eigenenergies were derived in Ref. [111].

1.3.3 Semiclassical quantisation

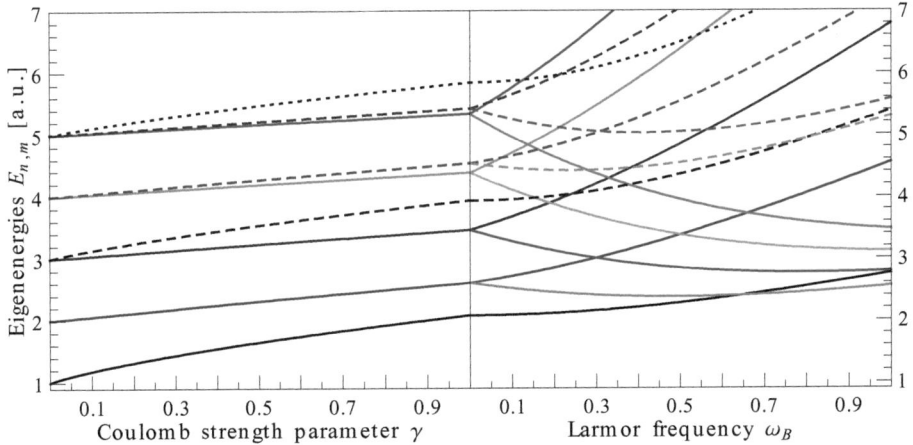

Figure 1.1: Low-lying eigenenergies of the relative motion H_{rel} for $\omega_0 = 1$. In the left plot we vary the Coulomb repulsion from non-interacting to full interaction. In the right plot we keep the full Coulomb interaction and vary the Larmor frequency up to 1 a.u. (in the case of InAs quantum dots this is equivalent to a magnetic field of approximately 1 Tesla). The results were obtained by numerically solving the semiclassical Bohr-Sommerfeld quantisation rule. The colours denote the angular momentum $m_r \in \{0, \pm 1, \pm 2, \pm 3, \pm 4\}$ (black, red, blue, green, magenta), while the line style represents the principal quantum number $n_r \in \{0, 1, 2\}$ (solid, dashed, dotted). The magnetic field lifts the degeneracy of positive (darker colour) and negative (lighter colour) angular momenta.

The one-dimensional radial Schrödinger equation (1.24) can also be solved approximately by application of Bohr-Sommerfeld quantisation with Langer modification [112]. The evaluation of the classical action is tedious, but feasible in terms of complete elliptic integrals of the first and third kind. The quantisation rule can then be solved numerically with minor effort and for different strengths of the interparticle interaction and the magnetic field, see Figure 1.1. The Coulomb repulsion lifts the degeneracy of the harmonic oscillator eigenstates, except that levels with equal principal quantum number n_r and equal absolute value angular momentum $|m_r| > 0$ are still degenerate. This degeneracy is lifted by the magnetic field and levels with negative angular momentum partially cross lower lying states. In Figure 1.1 the $|0,-1\rangle$ level (solid, red) has lower energy than the $|0,0\rangle$ state for $\omega_B > 0.65$. Thus, it becomes the ground state, which is the effect that accounts for the singlet-triplet flip of the ground state of Hooke's atom. For a comparison of the exact analytical and the semiclassical solutions see [113].

1.4 First results for Hooke's atom with anharmonic perturbation

In this section we show some results that confirm the experimental relevance of our model, discussed in Section 1.1. We will consider a few aspects of the anharmonic term and reproduce results from the literature for Hooke's atom. A detailed description of the numerical approach used here will be given in Chapter 2.

1.4.1 Spectrum

The spectrum of Hooke's atom emerges from the properly symmetrised four-dimensional harmonic oscillator spectrum described in Section 1.2.2 by turning on the interparticle interaction, see Figure 1.2 (left). The Coulomb repulsion between the electrons increases all eigenenergies. For the majority of states the effect of the Coulomb repulsion can be described by first-order perturbation theory, apparently clear from the linear behaviour of the levels. Some, but not all of the highly degenerate levels of the harmonic oscillator spectrum are lifted. States arising from the same harmonic oscillator eigenenergy form branches of close-lying energies, which can still be identified in the spectrum. Almost all states are strongly correlated linear combinations of several one-particle orbitals, corresponding to the separability in COM and relative coordinates.

1.4. First results for Hooke's atom with anharmonic perturbation

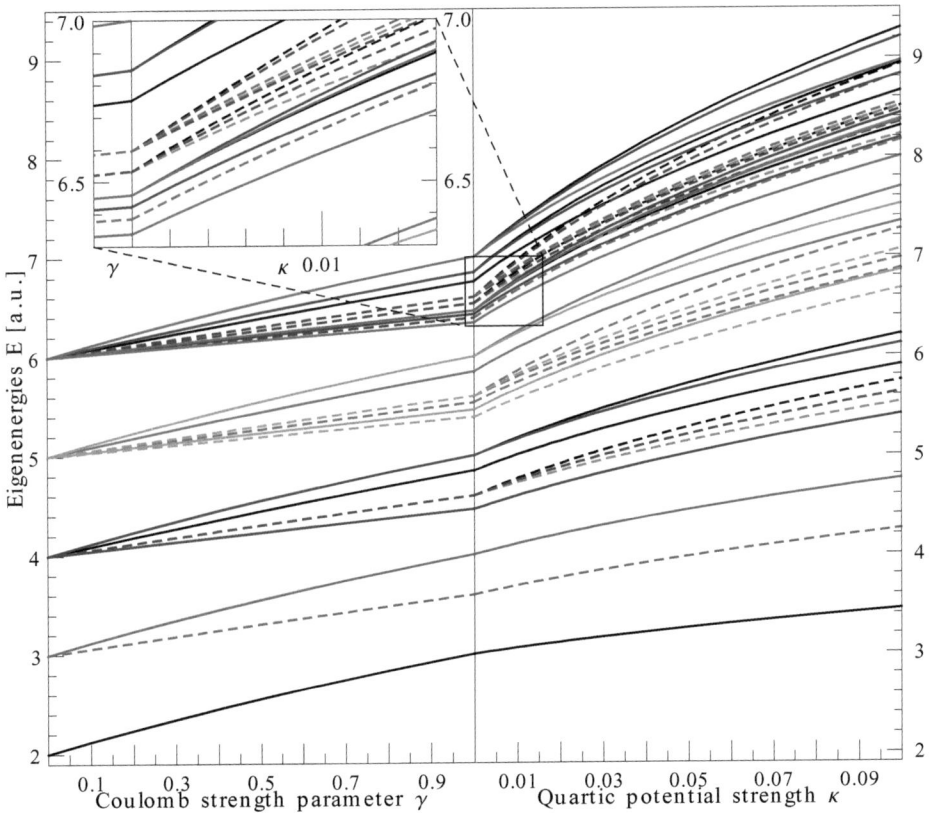

Figure 1.2: The spectrum of all states arising from harmonic oscillator energies $E \leq 6$ for $\omega = 1$. We show the effect of the Coulomb repulsion by varying the parameter γ in the left plot. The right plot shows the effect of the quartic potential by varying κ for $\gamma = 1$. The colours denote the angular momentum $m_r \in \{1, 2, 3, 4\}$ (red, blue, green, magenta). Zero angular momentum energies are depicted by black (grey) lines in case of even (odd) parity states. Line styles distinguish the spin symmetry between singlet (solid) and triplet (dashed) states. The inset shows the region of the branch emerging from harmonic oscillator energy $E = 6$, when the quartic potential lifts all degeneracies, except those for states with positive and negative equal angular momentum.

We will discuss this property in detail in Section 5.2.1 when we consider the applicability of Hund's first rule.

Turning on the quartic potential further increases the eigenenergies. A perturbative description is only applicable for very small values of the quartic potential strength κ. Within the branches of levels originally degenerate in the harmonic non-interacting model states are differently affected by the quartic potential. This leads to approaching levels and level repulsions (avoided crossings) for states belonging to the same symmetry class. For the lowest levels, as shown in Figure 1.2 (right and inset), the branches from equal harmonic oscillator energies are well separated from each other. For higher energies the branch-structure is lost, eigenenergies of different symmetry classes might cross each other and the density of avoided crossings increases considerably. The degeneracy of states that differ only in the sign of the angular momentum m_c or m_r is lifted with $\kappa \neq 0$. In particular states with vanishing angular momentum that differ only in parity will no longer be degenerate. See, for instance, the triplet states (dashed lines) emerging from the harmonic oscillator energy $E = 4$, where even and odd parity states $|0,1,0,-1\rangle^{\pm}$ (black and grey) grow apart. States differing only in the sign of the total angular momentum are still degenerate. This degeneracy can be lifted by a magnetic field.

1.4.2 Oscillator strengths

The assumption of dominantly parabolic confinement in the quantum dot is based on the experimental findings for the behaviour of the absorption spectra. Oscillator strengths for the planar quantum dot are defined by

$$f_{\text{fi}}^{(x)} = (E_{\text{final}} - E_{\text{initial}})|\langle \phi_{\text{final}}|x_1 + x_2|\phi_{\text{initial}}\rangle|^2, \qquad (1.28)$$

$$f_{\text{fi}}^{(y)} = (E_{\text{final}} - E_{\text{initial}})|\langle \phi_{\text{final}}|y_1 + y_2|\phi_{\text{initial}}\rangle|^2. \qquad (1.29)$$

They are a standardised tool for the theoretical prescription of the absorption spectrum in dipole approximation by the absorption cross section,

$$\begin{aligned}\sigma_{\text{abs}}(\omega) &= 4\pi^2 \omega |\boldsymbol{\pi}\langle\phi_{\text{final}}|\boldsymbol{r}_1 + \boldsymbol{r}_2|\phi_{\text{initial}}\rangle|^2 \delta(E_{\text{final}} - E_{\text{initial}} - \omega) \\ &= 4\pi^2 f_{\text{fi}}^{(x)} \delta(E_{\text{final}} - E_{\text{initial}} - \omega),\end{aligned} \qquad (1.30)$$

where we have assumed polarisation in x-direction in the second step. The oscillator strengths fulfil the Thomas-Reiche-Kuhn sum rule,

$$2 = \sum_{n} \left(f_{\text{ni}}^{(x)} + f_{\text{ni}}^{(y)} \right), \qquad (1.31)$$

where the sum goes over all final states and the result equals the number of electrons, $N = 2$. In dipole approximation the difference in total angular momentum between final and initial state is $\Delta m = \pm 1$.

In parabolic confinement the only oscillator strengths from the ground state $|0,0,0,0\rangle$ couple to the states $|0,\pm 1,0,0\rangle$ and equal unity. This mode, which is dominant in the experimental absorption spectra, only couples COM degrees of freedom and is called Kohn mode. Sako *et al.* analysed the oscillator strengths in a Gaussian potential quantum dot with few electrons [114]. They showed that the Kohn mode will be distributed to several modes, coupling also to the relative motion with growing anharmonicity. In our model this effect can also be observed, but even for purely quartic confinement ($\omega_0 = 0$, $\kappa = 1$) there is still one dominant mode, while the second largest value of an oscillator strength is below 0.01.

1.4.3 Influence of a magnetic field

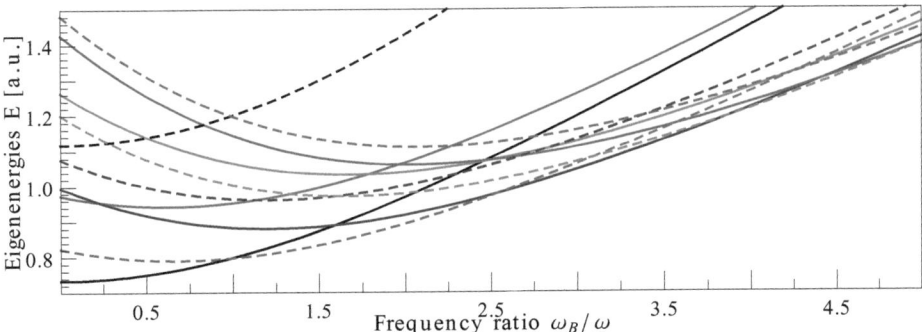

Figure 1.3: For Hooke's atom with quartic perturbation ($\kappa = 0.001$) we show the transition of the ground state between singlet and triplet states for application of a magnetic field. The harmonic confining frequency equals $\omega_0 = 0.1$ and the Larmor frequency ω_B is varied from 0 to 5. The colours denote the (negative) total angular momenta $m \in \{0, -1, -2, -3, -4\}$ (black, red, blue, green, magenta). Solid (dashed) lines are singlet (triplet) states. The zero-angular-momentum states have even parity. The transition between singlet and triplet states stems from the linear dependence of the eigenenergy of the relative motion on the magnetic field and the angular momentum (see Figure 1.1).

Most models for quantum dots include the application of a magnetic field perpendicular to the plane of the dot. The transition of singlet to triplet spin symmetry of the ground state has gained a lot attention from both experimental and theoretical physics. This effect is best seen for rather weak harmonic confinement, see Figure 1.3. We adapt the strength of the quartic perturbation to the weak harmonic confinement. The ratio, $\frac{1}{2}\omega_0^2/\kappa$ between harmonic and quartic potential at the distance of one natural length scale is kept constant. With these parameters we obtain the same behaviour as for the Hooke's atom [86], but with shifted magnetic field and energy scale.

As already mentioned the magnetic field lifts the degeneracy of corresponding states with positive and negative total angular momentum. As a consequence the oscillator strengths of the Kohn mode coupling the ground state to the states $|0, \pm 1, 0, 0\rangle$ differ according to the energy splitting.

Concerning the complexity of the system, which is directly linked to the separability and can be measured by several techniques, as we will show in Chapters 3 and 4, the magnetic field does not introduce new features. The effect on the harmonic confining potential can be incorporated by the choice of the harmonic frequency, while the Zeeman terms only introduce shifts of the energies. Therefore, we will omit application of a magnetic field in our further investigations and if not stated differently $\omega_B \equiv 0$. Nevertheless, we emphasise that magnetic fields at strengths that are easily accessible in the laboratory will render the dynamics mostly regular. The explicit strength of the magnetic field, naturally, depends strongly on the surrounding semiconductor expressed through the values of m^* and ϵ, see Appendix A.1.

1.5 Summary

In this chapter we presented an overview of the experimental results we consider fundamental for our numerical model. Our general Hamiltonian describes the dynamics of two electrons in a planar, radially symmetric confinement. The potential is dominantly harmonic with a quartic perturbation. The Coulomb interaction between the electrons is screened by the surrounding semiconductor, expressed through the relative permittivity and the effective mass incorporated in our choice of modified atomic units. Application of a magnetic field is also included. The spins of the electrons couple to give singlet and triplet states, distinguishable by the symmetry properties of the coordinate-dependent part of the wave function. We discuss symmetry properties of the underlying non-interacting harmonic model and show how the spectrum of the full model emerges from these highly degenerate levels. For purely harmonic confinement with Coulomb interaction our model reduces to Hooke's atom, the most common model for a two-electron quantum dot. The introduction of an additional anharmonic confinement maintains the key features corresponding to experimentally verified properties of quantum dots. In particular for low-lying states the spectral structure that was characteristic for the "table of artificial atoms" is preserved. The transition between singlet and triplet spin symmetry for the ground state of a quantum dot with applied magnetic field can be reproduced. Eventually, the dominance of the Kohn mode in the absorption spectrum was also found in the anharmonic case, although COM and relative motion become coupled. These first results were calculated with our numerical approach presented in detail in the following chapter.

Chapter 2

Numerical treatment

Quantum mechanical systems with exact analytical solutions are the exception and the application of approximate methods is required. Even few-body problems, with a small number of degrees of freedom belong to these cases. Semiclassical methods become extremely difficult, when the underlying classical dynamics is at least partially chaotic. The most prominent example, the helium atom, has challenged the physical community since the beginning of quantum mechanics [59]. Until today the *ab initio* calculation of highly doubly excited states of atomic helium can only be accomplished with immense computational efforts [115]. The dynamics of Hooke's atom is significantly simpler compared with the helium atom. Several approaches, which include exact diagonalisation [116, 117] and semiclassical methods [112, 101, 113] offered a detailed insight into the dynamics of Hooke's atom. A Hartree, a Hartree-Fock approximation and exact diagonalisation for Hooke's atom have been compared in Reference [12]. The introduction of an anharmonic potential, breaking the separability and as a consequence the regularity of the dynamics requires a sophisticated numerical treatment. Semiclassical arguments justify a planar approach to atomic helium, while it is intrinsic in the physical properties of quantum dots. Consequently a similar approach as the one used for planar atomic helium [118, 119] can be implemented to describe the planar anharmonic two-electron quantum dot model.

2.1 The Schrödinger equation in matrix representation

A planar two-electron quantum dot model with harmonic and quartic confinement and full Coulomb interaction between the electrons is described in modified atomic units by the Hamiltonian (1.7),

$$H = \sum_{j=1}^{2} \left[-\frac{1}{2}\nabla_j^2 + \frac{1}{2}\left(\omega_0^2 + \omega_B^2\right) r_j^2 + \kappa r_j^4 \right] + \omega_B L_z + \frac{\gamma}{r_{12}} + H_{\text{spin}}. \qquad (2.1)$$

The first step towards a numerical treatment is the transformation of the Hamiltonian (2.1) to COM and relative coordinates. With this we take advantage of the separability of the harmonic potential problem dominating our system. A representation in polar coordinates involves a term consisting of a trigonometric function of an angle in the quartic potential, while in the Cartesian representation a square root appears in the denominator of the Coulomb interaction. Our goal is to obtain a purely polynomial expression of finite degree in some coordinates and the according partial derivatives. We succeed by transforming to parabolic coordinates and multiplying the Schrödinger equation with the Jacobian of the complete coordinate transformation. Notice that this turns the solution of the Schrödinger equation from an eigenvalue problem into a generalised eigenvalue problem (GEVP). We obtain an equation of monomials, with degree no higher than twelve, in the coordinates and their partial derivatives, which can be expressed in harmonic oscillator creation- and annihilation operators. The radial symmetry in COM and relative motion will be exploited by considering circular operators, which are complex linear combinations of the former harmonic oscillator operators. The tedious manipulation of normal ordering the operator terms is achieved by symbolic calculation in a *Mathematica* code [118]. The representation of the GEVP in the Fock basis corresponding to the circular operators is the starting point for our numerical implementation. In the following we will give a detailed description of these analytical manipulations.

2.1.1 From the Hamiltonian to a generalised eigenvalue problem

Starting from the Cartesian coordinates x_j and y_j with $j \in \{1, 2\}$ for the two electrons we transform directly to the parabolic coordinates μ_k and ν_k with $k \in \{+, -\}$ according to COM $(+)$ and relative $(-)$ motion:

$$\mu_+ = \pm\sqrt{R_+ + x_+}, \qquad\qquad x_\pm = x_1 \pm x_2, \qquad (2.2)$$
$$\nu_+ = \operatorname{sgn}(y_+)\sqrt{R_+ - x_+}, \qquad y_\pm = y_1 \pm y_2, \qquad (2.3)$$
$$\mu_- = \pm\sqrt{R_- + x_-}, \qquad\qquad R_\pm = +\sqrt{x_\pm^2 + y_\pm^2}, \qquad (2.4)$$
$$\nu_- = \operatorname{sgn}(y_-)\sqrt{R_- - x_-}, \qquad \sqrt{g} = R_+ R_-, \qquad (2.5)$$

2.1. The Schrödinger equation in matrix representation

where $\sqrt{g}$ is the Jacobian of this transformation. Note that the parabolic coordinates $\mu_k \in \mathbb{R}$ can take negative values. This introduces a four-fold layer of the coordinate space, which has to be properly treated as described in the following Sections. We transform the kinetic energy T and the different terms of the potential V as well as the angular momentum L_z and obtain,

$$T = -\frac{1}{2}\left(\frac{1}{R_+}\partial^2_{\mu_+} + \frac{1}{R_+}\partial^2_{\nu_+} + \frac{1}{R_-}\partial^2_{\mu_-} + \frac{1}{R_-}\partial^2_{\nu_-}\right), \tag{2.6}$$

$$V_{\text{harm}} = \frac{1}{16}\left[(\mu_+^2 + \nu_+^2)^2 + (\mu_-^2 + \nu_-^2)^2\right], \tag{2.7}$$

$$V_{\text{quart}} = \frac{1}{256}\left[(\mu_+ + \nu_-)^2 + (\mu_- - \nu_+)^2\right]^2 \left[(\mu_+ - \nu_-)^2 + (\mu_- + \nu_+)^2\right]^2$$
$$+ \frac{1}{256}\left[(\mu_- - \mu_+)^2 + (\nu_- - \nu_+)^2\right]^2 \left[(\mu_- + \mu_+)^2 + (\nu_- + \nu_+)^2\right]^2, \tag{2.8}$$

$$V_{\text{int}} = \frac{2}{\mu_-^2 + \nu_-^2}, \tag{2.9}$$

$$L_z = -\frac{i}{2}\left(\mu_+ \partial_{\nu_+} - \nu_+ \partial_{\mu_+} + \mu_- \partial_{\nu_-} - \nu_- \partial_{\mu_-}\right). \tag{2.10}$$

After substituting these terms in the Hamiltonian (2.1) the stationary Schrödinger equation $H|\psi\rangle = E|\psi\rangle$ takes the form

$$\sqrt{g}\left[T + \left(\omega_0^2 + \omega_B^2\right)V_{\text{harm}} + \kappa V_{\text{quart}} + \gamma V_{\text{int}} + \omega_B L_z\right]|\psi\rangle = \sqrt{g}E|\psi\rangle, \tag{2.11}$$

which is obtained after multiplication by the Jacobian of the coordinate transformation. The multiplication is motivated by the wish to regularise the denominator in the interaction term, which now reads $\sqrt{g}\,V_{\text{int}} = \frac{1}{2}(\mu_+^2 + \nu_+^2)$. This term in particular and all the potential terms in general are now polynomials in the parabolic coordinates. The kinetic term and the angular momentum are polynomials of the parabolic coordinates and the corresponding partial derivatives. The solution of the Schrödinger equation is equivalent to the solution of a GEVP, $A|\psi\rangle = E\,B|\psi\rangle$, with formal operators A and B. In the following we will transform this formal equation into a matrix equation suitable for numerical implementation.

2.1.2 From coordinates to operators

The coordinates and partial derivatives are pairwise transformed to harmonic oscillator creation- and annihilation operators. We skip this intermediate step

Chapter 2. Numerical treatment

and transform directly to the circular operators,

$$a_1 = \frac{1}{2}\left(\mu_+ + \partial_{\mu_+} - i\nu_+ - i\partial_{\nu_+}\right), \quad a_1^\dagger = \frac{1}{2}\left(\mu_+ - \partial_{\mu_+} + i\nu_+ - i\partial_{\nu_+}\right), \quad (2.12)$$

$$a_2 = \frac{1}{2}\left(\mu_+ + \partial_{\mu_+} + i\nu_+ + i\partial_{\nu_+}\right), \quad a_2^\dagger = \frac{1}{2}\left(\mu_+ - \partial_{\mu_+} - i\nu_+ + i\partial_{\nu_+}\right), \quad (2.13)$$

$$a_3 = \frac{1}{2}\left(\mu_- + \partial_{\mu_-} - i\nu_- - i\partial_{\nu_-}\right), \quad a_3^\dagger = \frac{1}{2}\left(\mu_- - \partial_{\mu_-} + i\nu_- - i\partial_{\nu_-}\right), \quad (2.14)$$

$$a_4 = \frac{1}{2}\left(\mu_- + \partial_{\mu_-} + i\nu_- + i\partial_{\nu_-}\right), \quad a_4^\dagger = \frac{1}{2}\left(\mu_- - \partial_{\mu_-} - i\nu_- + i\partial_{\nu_-}\right). \quad (2.15)$$

They fulfil the usual bosonic commutation relations,

$$[a_i, a_j] = [a_i^\dagger, a_j^\dagger] = 0, \quad (2.16)$$

$$[a_i, a_j^\dagger] = \delta_{ij}. \quad (2.17)$$

Thus, we can associate a harmonic oscillator basis $\{|n_j\rangle\}$ to each pair of circular operators $a_j^\dagger$ and a_j for $j \in \{1, 2, 3, 4\}$:

$$|n_j\rangle = \frac{(a_j^\dagger)^{n_j}}{\sqrt{n_j!}}|0\rangle, \quad (2.18)$$

and define the corresponding number operators $N_j = a_j^\dagger a_j$ such that $N_j|n_j\rangle = n_j|n_j\rangle$. We now express each term of the GEVP (2.11) in these circular operators (2.12-2.15). The normal ordering of these terms is performed with a *Mathematica* code [118]. After normal ordering the interaction term consists of 5, the kinetic term of 13, the Jacobian of 25, the harmonic term of 275 and the quartic term of 2088 ordered monomials. The maximum degree of all monomials is twelve reached only for the quartic potential terms. For example the interaction term reads

$$\sqrt{g}\, V_{\text{int}} = \frac{1}{2}\left(1 + a_1^\dagger a_1 + a_2^\dagger a_2 + a_1^\dagger a_2^\dagger + a_1 a_2\right). \quad (2.19)$$

Before we can translate these monomials into selection rules and algebraic expressions of some quantum numbers we need to define the basis on which the operators act.

2.1.3 Symmetries and the numerical basis

The circular operators act on the Fock basis,

$$|n_1, n_2, n_3, n_4\rangle = |n_1\rangle \otimes |n_2\rangle \otimes |n_3\rangle \otimes |n_4\rangle, \quad (2.20)$$

which is the tensor product of the basis states defined by Equation (2.18). There are two restrictions to this general basis. First, the parabolic coordinate transformation introduced a four-fold layer of coordinate space, which now leads to redundancies in the Fock basis. Second, we wish to restrict the basis to one of the specific symmetry classes, as discussed in Section 1.2.2. Each symmetry class can be characterised by the total angular momentum perpendicular to the plane, the parity and the total spin.

Continuous symmetry: Angular momentum

The angular momentum operator has a very simple representation in circular operators,

$$L_z = \frac{1}{2}(N_1 - N_2 + N_3 - N_4) = \frac{1}{2}(L_+ + L_-), \qquad (2.21)$$

where we introduced the angular momentum operators

$$L_+ = i\left(\nu_+ \partial_{\mu_+} - \mu_+ \partial_{\nu_+}\right) = a_1^\dagger a_1 - a_2^\dagger a_2 = N_1 - N_2, \qquad (2.22)$$
$$L_- = i\left(\nu_- \partial_{\mu_-} - \mu_- \partial_{\nu_-}\right) = a_3^\dagger a_3 - a_4^\dagger a_4 = N_3 - N_4, \qquad (2.23)$$

on the subspaces of COM and relative motion in parabolic coordinates, respectively.

Unphysical symmetries

The unphysical redundancies in the Fock basis are correlated with the four-fold layer in the parabolic coordinate space. Translating the multilayer property into an algebraic representation in terms of operators allows us to discover the action on the basis vectors. Physically relevant is the quotient space $\mathbb{R}^4/K$, where K is the Klein four-group, represented by the identity and the following three operators:

- Π_+: $(\mu_+, \nu_+) \to (-\mu_+, -\nu_+)$, parity operator in the parabolic COM coordinates,

- Π_-: $(\mu_-, \nu_-) \to (-\mu_-, -\nu_-)$ parity operator in the parabolic relative coordinates,

- Π_p: parity operator in parabolic coordinates, $\Pi_p = \Pi_+ \circ \Pi_-$.

Chapter 2. Numerical treatment

We consider the action of these operators on wave functions in the parabolic coordinate space $\psi(\mu_+, \nu_+, \mu_-, \nu_-)$:

$$\Pi_+ \psi(\mu_+, \nu_+, \mu_-, \nu_-) = \psi(-\mu_+, -\nu_+, \mu_-, \nu_-) = e^{-i\pi L_+} \psi(\mu_+, \nu_+, \mu_-, \nu_-),$$
$$\Pi_- \psi(\mu_+, \nu_+, \mu_-, \nu_-) = \psi(\mu_+, \nu_+, -\mu_-, -\nu_-) = e^{-i\pi L_-} \psi(\mu_+, \nu_+, \mu_-, \nu_-),$$
$$\Pi_p \psi(\mu_+, \nu_+, \mu_-, \nu_-) = \psi(-\mu_+, -\nu_+, -\mu_-, -\nu_-) = e^{-i2\pi L_z} \psi(\mu_+, \nu_+, \mu_-, \nu_-).$$

The operator representation of the action can be carried forward to give the action on the Fock basis (2.20):

$$\Pi_+ |n_1, n_2, n_3, n_4\rangle = \lambda_+ |n_1, n_2, n_3, n_4\rangle = e^{-i\pi(n_1-n_2)} |n_1, n_2, n_3, n_4\rangle, \quad (2.24)$$
$$\Pi_- |n_1, n_2, n_3, n_4\rangle = \lambda_- |n_1, n_2, n_3, n_4\rangle = e^{-i\pi(n_3-n_4)} |n_1, n_2, n_3, n_4\rangle, \quad (2.25)$$
$$\Pi_p |n_1, n_2, n_3, n_4\rangle = \lambda_+ \lambda_- |n_1, n_2, n_3, n_4\rangle. \quad (2.26)$$

These operators must leave invariant any physically relevant basis. Thus, we exclude the unphysical basis states by imposing the following restrictions on the quantum numbers:

$$\lambda_+ \stackrel{!}{=} +1 \iff (n_1 - n_2)(\text{mod}\, 2) \equiv 0, \quad (2.27)$$
$$\lambda_- \stackrel{!}{=} +1 \iff (n_3 - n_4)(\text{mod}\, 2) \equiv 0. \quad (2.28)$$

As a consequence the total angular momentum, Equation (2.21), will be integer-valued, as we would have expected.

Discrete physical symmetries

Besides the angular momentum we also identify the symmetry class by spin symmetry and parity. The corresponding operators are characterised by their action on the Cartesian basis for two particles:

- Π_{12}: $(x_1, y_1, x_2, y_2) \to (x_2, y_2, x_1, y_1)$, exchange of the two electrons,
- Π_{xy}: $(x_1, y_1, x_2, y_2) \to (y_1, x_1, y_2, x_2)$, interchange of the Cartesian coordinates.

The action of these operators on a wave functions in the parabolic coordinate space $\psi(\mu_+, \nu_+, \mu_-, \nu_-)$ is given by

$$\Pi_{xy}\psi(\mu_+, \nu_+, \mu_-, \nu_-) = \psi\left(\frac{|\mu_+ + \nu_+|}{\sqrt{2}}, \text{sgn}(\mu_+)\frac{\mu_+ - \nu_+}{\sqrt{2}}, \frac{\mu_- + \nu_-}{\sqrt{2}}, \text{sgn}(\mu_-)\frac{\mu_- - \nu_-}{\sqrt{2}}\right),$$
$$\Pi_{12}\psi(\mu_+, \nu_+, \mu_-, \nu_-) = \psi(\mu_+, \nu_+, \text{sgn}(\nu_-)\nu_-, -\text{sgn}(\nu_-)\mu_-)$$
$$= e^{\pm i\pi L_-/2} \psi(\mu_+, \nu_+, \mu_-, \nu_-).$$

2.1. The Schrödinger equation in matrix representation

The multiplicity in the action of the two operators will be cancelled for the Fock basis. It is convenient to distinguish singlet from triplet states, as well as even from odd parity states considering the following actions on the Fock basis (2.20):

$$\Pi_{xy}|n_1, n_2, n_3, n_4\rangle = |n_2, n_1, n_4, n_3\rangle, \tag{2.29}$$

$$\Pi_{12}|n_1, n_2, n_3, n_4\rangle = \epsilon_s|n_1, n_2, n_3, n_4\rangle = e^{\pm i\pi(n_3-n_4)/2}|n_1, n_2, n_3, n_4\rangle. \tag{2.30}$$

The Fock basis is inherently an eigenbasis of the particle exchange operator Π_{12} and the spin symmetry of the state can be determined by:

$$(n_3 - n_4) \pmod{4} \equiv \begin{cases} 0 & \text{singlet}, \ \epsilon_s = +1, \\ 2 & \text{triplet}, \ \epsilon_s = -1. \end{cases} \tag{2.31}$$

Numerical basis

The parity operator does not commute with the total angular momentum $[\Pi_{xy}, L_z] \neq 0$. As discussed in Section 1.2.2 for states with vanishing angular momentum we build an eigenbasis for the parity operator Π_{xy}. For this purpose we need to symmetrise the eigenstates,

$$|n_1, n_2, n_3, n_4\rangle^{\epsilon_p} = \begin{cases} |n_1, n_2, n_3, n_4\rangle, & n_1 = n_2 \wedge n_3 = n_4, \\ \frac{1}{\sqrt{2}} \left(|n_1, n_2, n_3, n_4\rangle + \epsilon_p |n_2, n_1, n_4, n_3\rangle \right), & \text{otherwise}, \end{cases} \tag{2.32}$$

to obtain eigenvectors of the parity-operator with eigenvalue ϵ_p. In order to guarantee uniqueness of the basis representation we impose restrictions on the quantum numbers resulting in three cases:

1) $n_3 > n_4$; n_1, n_2 arbitrary,
2) $n_3 = n_4$ and $n_1 > n_2$,
3) $n_3 = n_4$ and $n_1 = n_2$, but then $|n_1 n_2 n_3 n_4\rangle^{\epsilon_p \equiv 1} = |n_1 n_2 n_3 n_4\rangle$.

For non-vanishing angular momentum we can use the same symmetrisation and restrict the basis such that $m = \frac{1}{2}|n_1 - n_2 + n_3 - n_4|$. Alternatively we do not further symmetrise the basis $|n_1, n_2, n_3, n_4\rangle$ for $m \neq 0$ and accept also negative values of m. Only the latter choice allows for a representation of the angular momentum term describing the effect of a magnetic field.

2.1.4 Matrix representation

For the sake of simplicity we will address the representation of the matrices A and B of the GEVP in the basis (2.20) without parity symmetrisation. The effect of the symmetrisation can be incorporated in a straightforward manner. For a general operator O consisting of monomials in the circular operators a_j and $a_j^\dagger$ we consider the infinite dimensional representation

$$O = \sum_{n,n'} |n'\rangle\langle n'|O|n\rangle\langle n|.$$

Here we have used the abbreviation $|n\rangle := |n_1, n_2, n_3, n_4\rangle$ and the sum over n obeys the restrictions (2.27) and (2.28). Each monomial in the operator representation of the operator O couples one basis element $|n\rangle$ to exactly one basis element $\langle n'|$ and defines a selection rule $\Delta n = \{\Delta n_1, \Delta n_2, \Delta n_3, \Delta n_4\}$. Thus, a matrix element $\langle n'|O|n\rangle$ is non-zero, if a selection rule is defined such that

$$\langle n'| = \langle n + \Delta n| = \langle n_1 + \Delta n_1, n_2 + \Delta n_2, n_3 + \Delta n_3, n_4 + \Delta n_4|.$$

As an example the interaction term (2.19) defines three selection rules,

$$(1 + a_1^\dagger a_1 + a_2^\dagger a_2) \leftrightarrow \Delta n = \{\ 0,\ 0,\ 0,\ 0\},$$
$$(a_1^\dagger a_2^\dagger) \leftrightarrow \Delta n = \{\ 1,\ 1,\ 0,\ 0\},$$
$$(a_1 a_2) \leftrightarrow \Delta n = \{-1, -1,\ 0,\ 0\}.$$

Knowing which matrix elements are non-zero we still need to evaluate their specific value. The established actions of the circular operators on the basis functions translate into algebraic expressions involving square roots of the integer-valued quantum numbers n_1, n_2, n_3 and n_4. For example the matrix element of the quartic potential operator for the selection rule $\Delta n = \{-1, 3, -1, -5\}$ reads

$$\langle n + \Delta n|\sqrt{g}\,V_{\text{quart}}|n\rangle = \sqrt{n_1(n_2+1)(n_2+2)(n_2+3)n_3(n_4-3)(n_4-2)(n_4-1)n_4}$$
$$\times \left(\tfrac{5}{128} + \tfrac{5}{256}(n_1 - 1) + \tfrac{5}{512}n_2\right).$$

In total the operator A defines 171 selection rules, while the operator B defines only nine selection rules. Due to the finite number of selection rules the resulting matrices are sparse and banded. The great advantage of this approach is that the matrix elements are known as exact algebraic expressions.

For the numerical implementation we need to truncate the infinite basis to some finite number n_{tot}. We will only consider basis states for which $n_1 + n_2 + n_3 + n_4 \leq n_{\text{base}}$ and thus control the basis size with the parameter n_{base}. Notice that the bandwidths of our finite matrices depend on the ordering of the basis. In order to optimise the bandwidth of the matrices we adapt a special procedure to construct our basis, which is described in detail in [120].

2.2 Numerical solution of the generalised eigenvalue problem

In the last Section we have turned the physical problem we address, Equation (2.1), into a finite dimensional GEVP. In a truncated basis representation only a fraction of eigenvectors and eigenvalues of the GEVP will be physically meaningful. In order to obtain a certain number of well-converged eigenvalues the size of the numerical basis needs to exceed this number considerably. From experience one can tell that the number of accurate eigenvalues will only be approximately one tenth of the basis dimension. Therefore, some effort has to be expended on the numerical implementation of the approach. Three main tasks need to be taken into account. First, the matrices have to be stored in a proper scheme to keep the amount of necessary main memory as low as possible and to guarantee short access times. Second, the algorithm to diagonalise the matrices needs to be efficient and highly accurate. Third, there is stringent necessity to provide tools to test the accuracy and physical relevance of the results.

2.2.1 Diagonalisation and storage

Hamiltonian operators are self-adjoint and therefore the matrix representations are hermitian or even symmetric for a real-valued representation. In our case both involved matrices A and B are real, symmetric, banded matrices. The bandwidth is reduced by the recursive construction of the basis. Furthermore, we can consider states of different symmetry classes separately, leading to a block structure of the matrices. Each block can be characterised by the quantum numbers $(|m|, \epsilon_s, \epsilon_p)$ for general m or (m, ϵ_s) for $|m| > 0$. The total memory usage of a matrix stored exploiting the specific structure is $n_{\text{tot}} \times n_{\text{band}} \times 16$ bytes for double precision float numbers.

Lanczos algorithm

For the partial diagonalisation of large matrices specific algorithms have been developed. A Krylov subspace method developed for banded, symmetric matrices – generally known as the Lanczos algorithm [121] – is particularly suitable for such a purpose. This method efficiently calculates the largest eigenvalues of the problem. We take advantage of this property by shifting the energy and solving the inverse problem, which leads to a well converged spectrum around the shifted energy [122]. In general three major steps need to be performed. First the matrix A will be decomposed in a lower band triangular matrix $A = LL^T$ by Cholesky decomposition. This makes the multiplication of the inverse matrix A^{-1} accessible by backward substitution. A tridiagonal matrix T is then constructed by recursively acting with $A^{-1}B$ on the basis vectors of the Krylov subspace. With each of these Lanczos steps the basis vectors of the Krylov subspace are simultaneously built. The total number of Lanczos steps generally equals the dimension of the Krylov subspace. In a final step T is diagonalised by standard diagonalisation techniques and the eigenvalues are obtained. The basis vectors of the Krylov subspace define the transformation matrix suitable to obtain the corresponding eigenvectors. The eigenvectors are orthonormal only with respect to a non-trivial metric. This metric is defined by the Jacobian, or overlap matrix B, such that for a general eigenvector $|\psi_n\rangle$

$$\langle\psi_m|B|\psi_n\rangle = \delta_{n,m}. \tag{2.33}$$

2.2.2 Coordinate scaling and convergence

In order to adapt our basis functions to the parameters of the system and to verify the convergence of the eigenvalues we introduce a scaling transformation. The unitary operator $P_\alpha = \exp[-\frac{i}{2}(\bm{r}\cdot\bm{p} + \bm{p}\cdot\bm{r})\log\alpha]$, with α a real scaling factor, scales the coordinates and momenta according to $\bm{r} \to \alpha\bm{r}$ and $\bm{p} \to \bm{p}/\alpha$. The physical properties are not changed by this transformation, in particular the correct eigenvalues do not depend on α,

$$\frac{dE_\alpha}{d\alpha} = \langle E_\alpha|\frac{dH}{d\alpha}|E_\alpha\rangle = 0, \tag{2.34}$$

following the Hellmann-Feynman theorem. The truncation of the basis introduces a dependence on α for the numerical eigenvalues and observables, see Figure 2.1. Therefore the expectation value $\langle E_\alpha|\frac{dH}{d\alpha}|E_\alpha\rangle$ will not anymore be exactly zero, but at least small for well-converged eigenvalues. Thus we can treat α as a variational parameter to determine the accuracy of our results. In Figure 2.1 we present eigenenergies E_α and expectation values of r_{rel} and of $|dH/d\alpha|$ depending on α. In the range where the eigenenergies and the expectation values are stationary with respect to α we refine our search and determine an optimised value of α. For such an optimal value of α we have analysed the numerical results and find at least three figures of the eigenenergies converged with the restriction $|dH/d\alpha| \leq 0.1$ (dashed line in Figure 2.1 a)). Notice that the optimal choice of α depends on the potential selected and the symmetry class considered. In general the accuracy of the eigenenergies is better than the accuracy of the expectation values. We corroborate the convergence of our results by comparing to literature data for particular choices of the parameters.

2.2. Numerical solution of the generalised eigenvalue problem

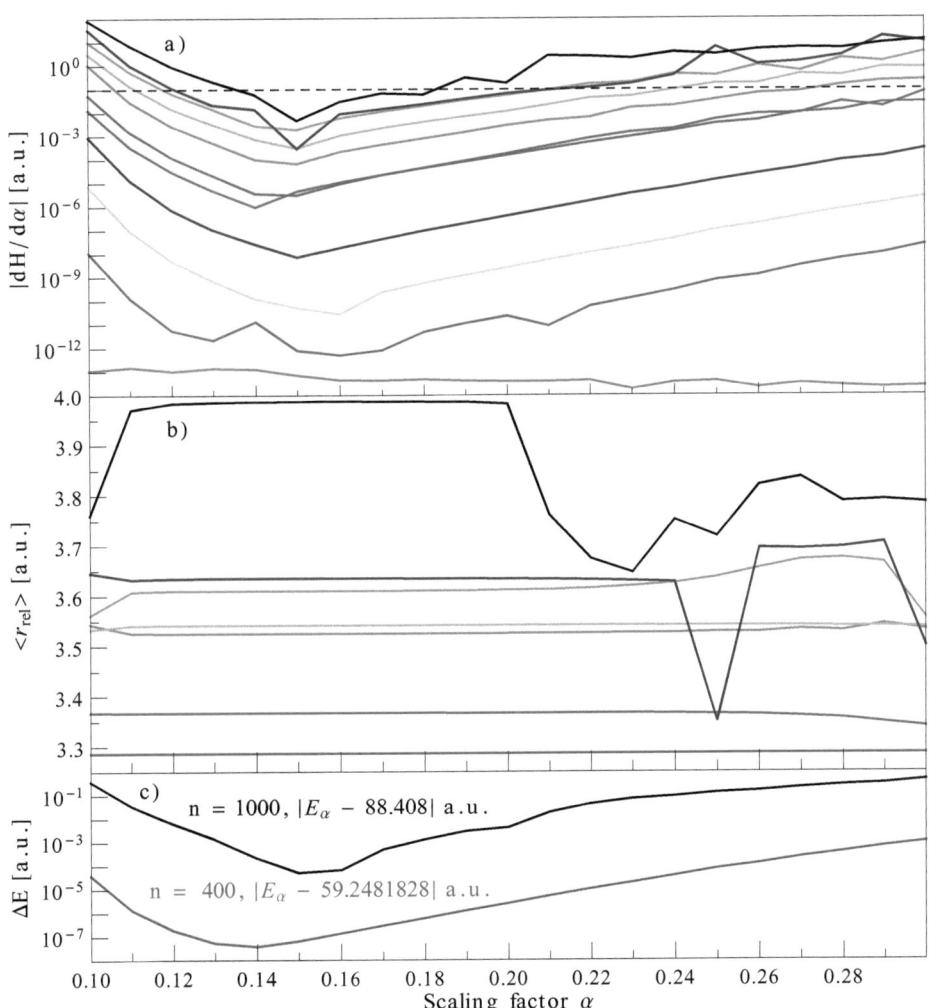

Figure 2.1: The dependence of the numerical results on the scaling factor α is shown for calculations with the parameters $\omega_0 = 1$, $\kappa = 0.1$, $\gamma = 1$, spin singlet symmetry, vanishing angular momentum and even parity. The total basis size is $n_{\text{tot}} = 10656$ and the bandwidth of the matrices is 1698. We have performed a diagonalisation with 3000 Lanczos steps for each value of $\alpha \in \{0.1, 0.11, .., 0.3\}$ and found more than 1000 roughly converged eigenvalues, respectively. In a) the value of the convergence parameter $|dH/d\alpha|$ is shown for every 100 states from the ground state ($n = 0$, grey) up to the 1000th excited state. The dashed line denotes our limit below which we consider eigenvalues to be converged for an optimised value of α. In b) we show the expectation values of the radial distance in the relative coordinates, starting from $n = 400$, evaluated as described in Section 2.3.2. The typical behaviour of two eigenenergies as function of α is illustrated in c).

2.2.3 Comparison to literature data

Our model can be regarded as one natural extension of the commonly used Hooke's atom model for quantum dots. To our knowledge there exist no literature data for the model including harmonic and quartic potential. Still we can compare our numerical results to the analytical results for Hooke's atom presented in Section 1.3.2. In addition, in the non-interacting case for purely quartic confinement the eigenenergies we obtain must be the sum of the eigenenergies of one-particle solutions in a planar quartic potential [88].

Analytical results for Hooke's atom

An explicit analytical solution, like the solution (1.25) for Hooke's atom, is an optimal testing ground for a numerical approach. We consider the explicit analytical solution with $m_r = 0$ and $n = 2$ according to (1.26) without magnetic field and obtain $\omega_0 = 1$ and $E_{\text{rel}} = 2$. This case appears for singlet spin symmetry and even parity for vanishing total angular momentum, $|n_c, 0, 0, 0\rangle^+$. For the COM motion the eigenenergy is simply $E_{\text{COM}} = 2n_c + 1$. We cannot only compare the eigenenergies $E = E_{\text{rel}} + E_{\text{COM}}$, but also other physical quantities derived from the coordinate representation. In particular the radial expectation value of the relative motion is given by

$$\langle r_{\text{rel}} \rangle = \frac{2(2 + \sqrt{2\pi})}{3 + \sqrt{2\pi}} = 1.636801341900272, \quad (2.35)$$

rounded to the last digit. The radial expectation value of the COM motion can be given by an integral expression[1], which is easily evaluated numerically to arbitrary precision. We present the according results for an optimal choice of $\alpha = 0.2$ and a rather small basis of dimension $n_{\text{tot}} = 6370$ in combination with the analytical results in Table 2.1. Notice, that the total number of states obtained for this calculation is 309, for which the last energy in Table 2.1 is the value with the worst accuracy accepted. Comparing to Table A.5 we find, that all states (252) up to energy $E = 27$ were obtained, while only a minor fraction of the spectrum could be evaluated above this value.

---

[1] $\langle R_{\text{COM}}(n_c) \rangle = \frac{1}{\sqrt{2}} \int_0^\infty dx \sqrt{x} \, [L_{n_c}(x)]^2 \, e^{-x}$

2.2. Numerical solution of the generalised eigenvalue problem

Table 2.1: Numerically and analytically calculated eigenenergies E and expectation values $\langle dH/d\alpha \rangle$, $\langle R_{\text{COM}} \rangle$ and $\langle r_{\text{rel}} \rangle$ for Hooke's atom. We compute the eigenenergies and the expectation values with a basis size of $n_{\text{tot}} = 6370$ for zero angular momentum, singlet spin and even parity. For this special choice of the parameters and symmetry class all eigenenergies and expectation values are known analytically. The eigenenergies analytically take exact odd integer values and $\langle r_{\text{rel}} \rangle = 1.636801341900272$, while the analytical expectation values of $\langle R_{\text{COM}} \rangle$ are given in the table.

E [a.u.]	$\langle dH/d\alpha \rangle$ [a.u.]	$\langle R_{\text{COM}} \rangle$ [a.u.] (numerical)	$\langle R_{\text{COM}} \rangle$ [a.u.] (analytical)	$\langle r_{\text{rel}} \rangle$ [a.u.] (numerical)
2.999 999 999 999 999	0.000 000 000 000 049	0.626 657 068 657 750	0.626 657 068 657 750	1.636 801 341 900 274
5.000 000 000 000 005	-0.000 000 000 000 008	1.096 649 870 151 057	1.096 649 870 151 063	1.636 801 341 900 264
7.000 000 000 000 039	-0.000 000 000 000 012	1.419 769 921 177 681	1.419 769 921 177 715	1.636 801 341 900 233
9.000 000 000 000 684	-0.000 000 000 000 206	1.681 692 992 842 792	1.681 692 992 843 259	1.636 801 341 899 816
11.000 000 000 001 018	0.000 000 000 000 223	1.907 854 079 694 065	1.907 854 079 694 655	1.636 801 341 899 770
13.000 000 000 000 819	0.000 000 000 000 113	2.109 832 797 669 925	2.109 832 797 670 389	1.636 801 341 899 910
15.000 000 000 000 076	-0.000 000 000 000 100	2.294 050 646 216 108	2.294 050 646 216 114	1.636 801 341 900 266
17.000 000 000 000 028	0.000 000 000 000 184	2.464 507 391 773 477	2.464 507 391 773 425	1.636 801 341 900 298
19.000 000 000 000 298	0.000 000 000 000 050 815	2.623 894 283 580 835	2.623 894 283 588 085	1.636 801 341 895 081
21.000 000 000 132 665	-0.000 000 000 022 123 264	2.774 124 916 536 298	2.774 124 922 956 349	1.636 801 338 158 184
23.000 000 005 786 998	0.000 000 000 768 108 903	2.916 617 106 250 236	2.916 617 677 766 815	1.636 801 016 875 796
25.000 000 010 704 806	-0.000 000 002 561 041 945	3.052 460 181 107 163	3.052 458 493 794 680	1.636 802 260 485 613
27.000 000 121 766 810	0.000 000 304 331 105 631	3.182 573 077 654 419	3.182 500 734 139 942	1.636 838 163 687 941
29.000 001 160 891 170	-0.000 000 115 325 875 665	3.307 270 472 895 011	3.307 429 480 650 727	1.636 722 556 978 399
31.000 002 459 204 911	0.000 000 453 485 791 883	3.427 756 436 953 774	3.427 804 635 093 639	1.636 775 103 185 396
32.999 974 115 691 302	-0.000 000 709 199 538 930	3.548 166 414 577 438	3.544 090 785 080 888	1.638 653 874 804 282
35.000 074 767 878 694	-0.001 304 748 848 729	3.658 326 733 913 608	3.656 678 487 565 035	1.637 404 037 276 759
37.000 475 686 981 808	-0.003 970 214 278 542	3.782 104 880 243 939	3.765 899 804 871 662	1.643 074 413 780 297
39.002 630 249 149 078	-0.058 636 365 021 995	3.918 894 728 959 541	3.872 039 883 745 950	1.653 311 492 871 477

Chapter 2. Numerical treatment

Numerical results for the two-dimensional quartic oscillator

Our code is optimised for calculating interacting electrons in a harmonic confinement with a quartic perturbation, showing only small deviations from the separability in COM and relative coordinates. Although for the purely quartic potential this separability is lost very accurate results are obtained, see Table 2.2. We will discuss properties of the quartic oscillator concerning the validity of Hund's first rule in Section 5.3. The agreement with the results taken from Ref. [88] enables us to determine the quantum numbers of the states obtained with our code by direct comparison with these results.

Table 2.2: Our numerical results are compared to the one-particle solutions taken from [88] for the planar quartic oscillator ($\kappa = 1$). The sum of two one-particle levels equals the corresponding two-particle solution for independent particles ($\gamma = 0$). We choose eigenenergies E of singlet states with even parity for both electrons in the same state. For the results presented here we compare the values of calculations for different values of α in order to find the most accurate result obtainable with our code. The basis size is $n_{\text{tot}} = 6370$ and $n_{\text{tot}} = 19360$ for $m = 0$ and $m = 2$, respectively.

$\|n_1, m_1, n_2, m_2\rangle^{\epsilon_p, \epsilon_s}_{\text{idp}}$	$E_1 = E_2$ [a.u.] [88]	$E_1 + E_2$ [a.u.]	E [a.u.] (numerical)
$\|0,0,0,0\rangle^{+,+}_{\text{idp}}$	1.477149754	2.954299508	2.95429950715599
$\|1,0,1,0\rangle^{+,+}_{\text{idp}}$	6.003386083	12.006772166	12.0067721666167
$\|2,0,2,0\rangle^{+,+}_{\text{idp}}$	11.802433595	23.604867190	23.604867190269
$\|3,0,3,0\rangle^{+,+}_{\text{idp}}$	18.458818701	36.917637402	36.917637408155
$\|4,0,4,0\rangle^{+,+}_{\text{idp}}$	25.791792370	51.583584740	51.5835847572
$\|0,1,0,1\rangle^{+,+}_{\text{idp}}$	3.398150176	6.796300352	6.7963003520554
$\|1,1,1,1\rangle^{+,+}_{\text{idp}}$	8.700453815	17.400907630	17.40090762791
$\|2,1,2,1\rangle^{+,+}_{\text{idp}}$	14.977808375	29.955616750	29.95561674464
$\|3,1,3,1\rangle^{+,+}_{\text{idp}}$	21.999601031	43.999202062	43.999202069
$\|4,1,4,1\rangle^{+,+}_{\text{idp}}$	29.634879565	59.269759130	59.269759113

2.3 Observables and details of the system

The diagonalisation algorithm provides us not only with eigenenergies, but also with very accurate eigenstates as linear combinations of the basis states, Equation (2.32). We will briefly discuss some physical properties of the system that we can obtain from the eigenstates and their numerical representation.

2.3. Observables and details of the system

2.3.1 Expectation values of the potentials

By construction of our approach the different terms in the Hamiltonian can be considered separately, Equation (2.6) - (2.9). We can thus evaluate the matrix elements of the potentials V_{harm}, V_{quart}, V_{int} and the kinetic part T, individually.

Virial theorem

It is easily shown that a generalised virial theorem is satisfied,

$$2\langle T \rangle = 2\,\omega_0^2\,\langle V_{\text{harm}} \rangle + 4\,\kappa\,\langle V_{\text{quart}} \rangle - \gamma\,\langle V_{\text{int}} \rangle\,,$$

where $\langle \cdot \rangle$ denotes the expectation value in an eigenstate. This can be transformed into an equivalent expression for the eigenenergy E depending only on the potential terms of the system:

$$E = 2\,\omega_0^2\,\langle V_{\text{harm}} \rangle + 3\,\kappa\,\langle V_{\text{quart}} \rangle + \frac{1}{2}\gamma\,\langle V_{\text{int}} \rangle. \tag{2.36}$$

This result can also be used as a convergence test of our approach. Indeed, our numerical results agree with the virial theorem (2.36) up to the full accuracy of the eigenenergies.

Perturbative corrections

It is possible to calculate the matrix elements for potentials that are not part of the applied Hamiltonian. For example one can calculate the expectation value $\langle V_{\text{int}} \rangle$ for a non-interacting model and obtain the energy correction in first order perturbation theory for the Coulomb interaction. Additionally we can evaluate the non-diagonal matrix elements of the potentials in order to calculate higher orders of perturbation theory. An application will be the evaluation of the time-dependent quantum fidelity from purely static properties of our system in Section 4.3.1.

2.3.2 Expectation values and oscillator strengths

Physical observables $\langle O \rangle$ frequently reveal basic properties of quantum states. We evaluate the matrix elements by transforming the product operator BO into a matrix in the numerical basis. Notice, that the multiplication with the overlap matrix B is needed in order to account for the non-trivial scalar product of the eigenstates. We consider the following operators O:

- $\frac{1}{2}R_+$ the distance in COM motion,

Chapter 2. Numerical treatment

- R_- the distance in relative motion,
- r_j^2 the squared distance of the electron j,
- $(x_1 + x_2)$ for the oscillator strength $f_{\text{fi}}^{(x)}$,
- $(y_1 + y_2)$ for the oscillator strength $f_{\text{fi}}^{(y)}$,
- R_+^2 the squared distance in COM motion,
- R_-^2 the squared distance in relative motion.

The latter two operators are only calculated to estimate the cosine of the angle between the electron radii φ_{12} by

$$\cos \varphi_{12} \approx \frac{\langle \mathbf{r}_1 \cdot \mathbf{r}_2 \rangle}{\langle |\mathbf{r}_1| \rangle \langle |\mathbf{r}_2| \rangle} \approx \frac{\langle R_+^2 \rangle - \langle R_-^2 \rangle}{4\sqrt{\langle \mathbf{r}_1{}^2 \rangle}\sqrt{\langle \mathbf{r}_2{}^2 \rangle}}.$$

The procedure to evaluate the observables is similar to the one for evaluating the Hamiltonian matrix A and the overlap matrix B presented above. We obtain 84 selection rules for the radial expectation values, while there are 31 rules for the oscillator strengths. The matrix of the oscillator strengths couples angular momenta with $\Delta m = \pm 1$. Therefore, this matrix is not stored block-wise for each angular momentum, but as one matrix over the range of all angular momenta involved.

2.3.3 Densities of the states

Detailed information about the localisation of the states can be obtained by visualising the probability density in coordinate space. The configuration space of our approach is four-dimensional, such that it becomes necessary to integrate over at least two degrees of freedom to obtain a representable density. Several possibilities are implemented in our approach: the projection on the COM degrees of freedom, the projection on the relative degrees of freedom, the projection on the two one-particle radial distances and the projection onto the plane of the two radial distances in COM and relative coordinates. The latter contains the most information for most cases originated in the physical structure of separability.

In the polar coordinates associated to the parabolic coordinates (2.2) - (2.5) the basis functions can be represented in coordinate space with identification of the quantum numbers:

$$M = n_1 - n_2, \quad L = n_3 - n_4, \quad N = \min(n_1, n_2), \quad K = \min(n_3, n_4),$$

as follows [118]:

$$\varphi_n(r_+, \phi_+, r_-, \phi_-) = \langle n_1, n_2, n_3, n_4 \mid r_+, \phi_+, r_-, \phi_- \rangle =$$
$$\mathcal{N} \, r_+^{|M|} \, r_-^{|K|} \, L_N^{|M|}\left(r_+^2\right) L_K^{|L|}\left(r_-^2\right) e^{-\frac{1}{2}(r_+^2 + r_-^2)} e^{-\mathrm{i}(M\phi_+ + L\phi_-)},$$

where $L_n^k(x)$ are associated Laguerre polynomials and $\mathcal{N}$ is a normalisation constant:

$$\mathcal{N} = (-1)^{N+K} \frac{2}{\pi} \sqrt{\frac{N! \, K!}{(|M|+N)! \, (|L|+K)!}}.$$

Density plots are calculated by integrating over the angles in COM and relative coordinates, where the appropriate coordinate transformation is $r_+^2 = 4 R_{\text{COM}}$ and $r_-^2 = 2 r_{\text{rel}}$. For a general vector $|\psi\rangle = \sum_n a_n |n\rangle$, where the symmetrisation (2.32) is incorporated in the coefficients a_n, we plot the following function:

$$\sum_n \sum_{n'} a_n^* a_{n'} \int_0^\pi d\phi_+ \int_0^\pi d\phi_- \, R_{\text{COM}} \, r_{\text{rel}}$$
$$\times \varphi_n(\sqrt{4 R_{\text{COM}}}, \phi_+, \sqrt{4 r_{\text{rel}}}, \phi_-)^* \, \varphi_{n'}(\sqrt{4 R_{\text{COM}}}, \phi_+, \sqrt{4 r_{\text{rel}}}, \phi_-).$$

2.3.4 Angular momentum quantum numbers

The angular momentum quantum numbers in COM and relative coordinates, as introduced in (1.13) are good quantum numbers as long as separation in COM and relative motion is possible. It is preferable to obtain these quantum numbers as part of the basis representation $|\psi\rangle = |n_c, m_c, n_r, m_r\rangle^{\epsilon_p}$ for any state automatically. Furthermore, we find that, even in absence of the separability, these quantum numbers, though only approximate, give reasonable information on the localisation of the states. We distinguish between states that are parity symmetrised and those that are not. For a state with parity symmetrisation only the squares m_c^2 and m_r^2 are good quantum numbers. We evaluate them by considering the expectation values of the squared angular momentum operators $|m_c| = \sqrt{\langle\psi|L_+^2|\psi\rangle}$ and $|m_r| = \sqrt{\langle\psi|L_-^2|\psi\rangle}$, see (2.21). For a state without parity symmetrisation both angular momentum quantum numbers are good quantum numbers and we can directly evaluate $m_c = \langle\psi|L_+|\psi\rangle$ and $m_r = \langle\psi|L_-|\psi\rangle$.

2.4 Classical model

Apart from the quantum mechanical approach explained so far, we also consider the underlying classical dynamics of the system. Our system has four degrees of freedom and therefore an eight-dimensional phase space, which is effectively reduced to six dimensions by conservation of energy and total angular momentum. For $\gamma = 0$ the motion is regular, since the system separates in independent particles and for each particle the angular momentum and the energy is preserved. Following a similar argument for $\kappa = 0$ due to the separation in COM and relative coordinates the motion is also regular. In these cases the motion is bound to regular tori with a fundamental frequency associated to each constant of motion. In all other cases the dynamics is complex and we apply numerical methods to solve the equations of motion.

2.4.1 Description and time evolution

The classical model is expressed through the Hamilton function

$$H_{\text{class}} = T_{\text{class}} + \omega^2 V_{\text{harm}} + \kappa V_{\text{quart}} + \gamma V_{\text{int}}, \qquad (2.37)$$

where we do not consider a magnetic field and the potentials are to be understood as classical quantities. We express the single terms in Cartesian coordinates in COM and relative motion and adopt modified atomic units:

$$T_{\text{class}} = \frac{1}{2}(p_{x_c}^2 + p_{y_c}^2 + p_{x_r}^2 + p_{y_r}^2),$$

$$V_{\text{harm}} = \frac{1}{2}(x_c^2 + y_c^2 + x_r^2 + y_r^2)$$

$$V_{\text{quart}} = \frac{1}{2}\left[(x_c^2 + y_c^2 + x_r^2 + y_r^2)^2 + 4(x_c x_r + y_c y_r)^2\right]$$

$$V_{\text{int}} = \frac{1}{\sqrt{2(x_r^2 + y_r^2)}}.$$

The classical momenta are $p_j = \dot{q}_j$ with $q_j \in \{x_c, y_c, x_r, y_r\}$, where we implicitly introduced the four-dimensional vectors $\vec{p}$ and $\vec{q}$. It would be possible to eliminate the total angular momentum from the equations and obtain an effectively six-dimensional problem. Nevertheless, there are advantages in keeping the Cartesian representation. We only need to evaluate simple algebraic operations in the potentials and the preservation of the total angular momentum is a check for the accuracy of our code. Starting point for a numerical treatment of the classical motion are Hamilton's equations of motion, which can be written in a very compact form revealing the symplectic structure

$$\frac{d}{dt}\begin{pmatrix}\vec{q}\\\vec{p}\end{pmatrix} = J\begin{pmatrix}\vec{\nabla}_q\\\vec{\nabla}_p\end{pmatrix}H_{\text{class}}, \text{ with } J = \begin{pmatrix}0 & \mathbb{1}\\-\mathbb{1} & 0\end{pmatrix}. \qquad (2.38)$$

2.4. Classical model

It is advisable to choose a numerical method that supports this symplectic structure of the equation [123]. Except for numerical errors the energy and the angular momentum are automatically preserved by such a method. The very simple leapfrog algorithm is particularly efficient and therefore commonly applied for the calculation of dynamical systems. The time step δt is constant and the initial values need to be taken at $\vec{q}_0 = \vec{q}(t=0)$ and $\vec{p}_0 = \vec{p}(t=\delta t/2)$. For the coordinates and the momenta an explicit Euler method is applied:

$$\vec{q}_n = \vec{q}_{n-1} + \vec{p}_{n-1}\delta t, \qquad (2.39)$$

$$\vec{p}_n = \vec{p}_{n-1} + \vec{F}(\vec{q}_n)\delta t, \qquad (2.40)$$

where $\vec{F}(\vec{q}) = -\vec{\nabla}_q H_{\text{class}}$ depends only on the coordinates. Additionally we applied a sophisticated fourth order Runge-Kutta-Nyström algorithm [123] in order to check the accuracy of our method. The energy and the angular momentum is conserved with a relative error of the order 10^{-5} or better.

2.4.2 Frequency analysis

The precise presentation of the results obtained by the time evolution of initial configurations is a non-trivial task. A very convenient tool, Poincaré's surface of section can not be used, since it is restricted to an effectively three-dimensional phase space. A powerful alternative is the analysis of the dominant frequencies of a trajectory [124]. These frequencies coincide with the fundamental frequencies of the regular motion considered above. For non-regular dynamics the dominant frequencies can give an insight into the rich structure of the classical phase space. Several investigations of high-dimensional systems have been performed by frequency analysis, the Stark-Quadratic-Zeeman problem [125], the two-dimensional standard map [126, 127], different multidimensional systems [124] and the study of the stability of the solar system [67].

The time evolution of an initial condition results in a sequence of $\{\vec{q}_n, \vec{p}_n\}_n$ with $n \in \{n_{\text{initial}}, n_{\text{final}}\}$. For each degree of freedom j we consider the combination $\tilde{t}_{\tilde{n},j} = q_{\tilde{n},j} + i p_{\tilde{n},j}$ on a sub-interval $\tilde{n} \in \{\tilde{n}_{\text{initial}}, \tilde{n}_{\text{final}}\} \subset \{n_{\text{initial}}, n_{\text{final}}\}$. The equidistant sub-interval is equivalent to a time interval $T = (\tilde{n}_{\text{final}} - \tilde{n}_{\text{initial}})\Delta T$. The resolution in time ΔT determines the range of frequencies obtained and can be any multiple of the time step δt of the time evolution. Then we multiply with the widely used Hanning filter in order to avoid effects at the edges of the interval,

$$t_{\tilde{n},j} = \left(1 - \cos\left[2\pi(\tilde{n} - \tilde{n}_{\text{initial}})\frac{\Delta T}{T}\right]\right)\tilde{t}_{\tilde{n},j}. \qquad (2.41)$$

On this sequence we perform a discrete Fourier analysis and obtain a sequence of frequencies in the range $[-\frac{1}{2\Delta T}, \frac{1}{2\Delta T}]$ with an accuracy $\Delta f = 1/T$. We are interested in the frequency with the highest weight in the Fourier analysis, which we refer to as the dominant frequency. With a properly chosen time interval T it is possible to localise the correct region, where the dominant frequency can be found via the highest value of the frequency sequence. The exact frequency is evaluated in the vicinity of this, and the two neighbouring values, by an iterative scheme (golden section search). In each iteration the Fourier transform is obtained by straightforward quadrature of the Fourier integral. The gain in accuracy is usually several orders of magnitude compared to Δf.

2.5 Summary

Exploiting the separability properties of the system a proper coordinate transformation was introduced and the Schrödinger equation was transformed to a generalised eigenvalue problem consisting purely of polynomial expressions in coordinates and partial derivatives. Translating into harmonic oscillator operators led to a matrix representation with algebraic expressions in four integer-valued quantum numbers. The only approximation is the truncation of the basis to make the matrices finite and thus numerically implementable. Thus, the solution of the Schrödinger equation has been reduced to the solution of a generalised eigenvalue problem with sparse banded symmetric matrices with exact entries. A Lanczos algorithm is used to partially diagonalise the matrices. The accuracy of our results has been shown by comparison to the literature and an analysis based on a variational parameter scaling the basis functions. The determination of physical observables has been explained. Finally a classical approach including the method of frequency analysis has been described.

Chapter 3
Effects of the anharmonicity

In this chapter we investigate the effect of the anharmonicity on the complexity of the system. For that purpose we rely on tools that are related to the universal predictions of random matrix theory (RMT) [57]. These include the nearest-neighbour-spacing distribution $P(s)$ and the distribution $P_{\mathrm{ac}}(c)$ of the energy gaps c of the avoided crossings which appear by slow variation of the parameter κ. While the universality of RMT predictions has been confirmed by several experimental, semiclassical and numerical results for systems with underlying chaotic classical dynamics, its implications for systems with mixed regular-chaotic phase space – such as our system – are still subject of research and not entirely understood [79]. The term mixed regular-chaotic refers to classical phase space but is nonetheless used to characterise quantum systems. For these systems the deviations from the properties of purely chaotic systems can be explained by semiclassical arguments. We start this chapter with a brief description of some foundations of the theory of chaos and complexity, which are required for the understanding of our results. In particular we review the notions of classical chaos, semiclassical quantisation in chaotic dynamics and random matrix theory.

Classical chaos

Analytical considerations on chaotic systems go back to the 19th century, when Poincaré found the dynamics of the system moon-earth-sun to be non-integrable. A significant increase in the interest in such systems arose in the 1960's, when computing power was, for the first time, capable of calculating a great number of trajectories with high accuracy. Chaotic dynamics is an effect of non-linear dynamical systems and besides the already mentioned three-body problem with $\frac{1}{r}$-potentials, other famous examples from various fields came into the focus of research. For example the Lorenz model inspired from meteorology [68], the logistic map, describing a very simplified population model in the balance of reproduction and starvation [128], or the standard map as a model of a kicked rotor [129], to name only a few. Classical systems perform predictable motion, usually governed by the analytically exact equations of motion. Chaotic dynamics is at the same time random in the sense that close-lying trajectories may behave significantly different after some time, summarised in the property of sensitivity to initial conditions. Sophisticated mathematical investigations on this topic have led to several equivalent exact definitions of chaotic motion, but the conceptual approach presented appears to be sufficient for our purpose [130].

A system is regular, if the number of degrees of freedom, n, is accompanied by the same number of independent constants of motion in involution. For each degree of freedom a pair of action-angle variables can then be found by integration and the motion is restricted to an n-dimensional torus [131]. Nearly-integrable systems, the dynamics of which is chaotic due to small perturbations of a regular system, are of particular interest for investigations. The application of classical perturbation theory to these systems is governed by the problem of *small divisors* [132]. A very general result resolves this problem, the Kolmogorov-Arnold-Moser (KAM) theorem [133]. As a consequence some of the regular tori will be deformed, while others break up and form a "chaotic sea". This is exactly the situation of a mixed regular-chaotic phase space, where regular orbits and chaotic trajectories coexist. Nevertheless, for a large set of initial conditions the motion continues to be quasi-periodic as a consequence of the KAM theorem.

Semiclassical quantisation in chaotic dynamics

Semiclassical quantisation is an important foundation for many applications in modern quantum physics. The central idea is to discretise the classical action, intrinsically available for regular motion as described in the previous paragraph, in multiples of the Planck's constant $\hbar$. Exact quantisation of the non-relativistic hydrogen atom can be achieved with this approach [131]. The formal analogy of the classical dynamics of atomic helium to the gravitational three-body problem illustrates the challenges posed for semiclassical quantisation in this case [59]. On the one hand these obstacles encouraged the development of the alternative approaches by Schrödinger and Heisenberg to wave mechanics and a modern picture of quantum mechanics. On the other hand semiclassical theory was further developed along the years and for instance the quantisation of the atomic helium ground state and also of excited states was achieved. A generalisation of Feynman's path-integral formulation of quantum mechanics for semiclassical propagators resulted in Gutzwiller's trace formula describing the density of states by quantisation of periodic orbits [58]. It was shown that the density of states can be decomposed into a smooth and an oscillatory component [134]. The smooth part can be associated with periodic orbits of length zero reflecting properties of the available energy surface. In order to describe the oscillating component of the level density of a complex system with semiclassical methods a very precise knowledge of the periodic orbits is obligatory. This can in general only be achieved for low dimensional systems with effectively two degrees of freedom, e.g hydrogen in a magnetic field or one-particle billiards. In the latter case the smooth part of the level density can be approximated by Weyl's law [135] with a constant term depending only on the surface of the billiard and a term with a $\sqrt{E}^{-1}$ behaviour depending on the boundary length. Billiard systems play a crucial role in the investigation of chaotic systems, since the quantum properties can be modelled with microwaves. The formal analogy of the Schrödinger and the Helmholtz equation allows for a feasible experimental realisation of such systems providing full accessibility of eigenenergies and eigenfunctions [78].

Random matrix theory

While semiclassical quantisation attempts to give an exact characterisation of each quantised level, RMT aims at a statistical description of the spectral properties [136]. A model Hamiltonian is arranged in a way that the basic symmetry structures are preserved, but the components of the matrix representation are independent random variables. Three major symmetry classes are able to characterise most of the relevant physical systems, namely the Gaussian unitary ensemble, the Gaussian orthogonal ensemble (GOE) and the Gaussian symplectic ensemble. The ensembles are all called Gaussian, because it can be shown that for systems invariant under the according symmetry group (unitary, orthogonal, symplectic) they are most generally represented by Gaussian distributed random variables [137, 138]. By far the most common class is the GOE, which also applies to our model, since it is symmetric under time-reversal and under rotations, furthermore the Hamiltonian is real symmetric. The subtlety of the symmetry and their corresponding ensembles, which is fundamental for random matrix theory, will not be further discussed here.

Originally RMT had been applied to the complex spectra of nuclei [139], based on early ideas of Wigner [140]. But the great success of RMT stems from a result that showed to be widely applicable: the Bohigas-Gianonni-Schmit conjecture [57]. It predicts statistical similarity between RMT spectra and the energy levels of generic systems with underlying classically chaotic dynamics. In particular the distribution of energy gaps between nearest-neighbour levels $P(s)$ is claimed to show a universal behaviour. Though not formally proven this conjecture holds for an overwhelming number of experimental and numerical studies of systems with classical analogs with purely chaotic dynamics. A system showing this property is said to exhibit quantum chaos, also if no underlying classical system exists. For the relevant GOE the nearest-neighbour spacing distribution is predicted to obey the well-known Wigner distribution

$$P_{\text{Wigner}}(s) = \frac{\pi}{2} s \, e^{-\pi s^2/4}, \qquad (3.1)$$

while the spectra of regular systems follow the Poisson distribution

$$P_{\text{Poisson}}(s) = e^{-s}. \qquad (3.2)$$

Both distributions, (3.1) and (3.2), are the limiting cases for the interpolating distributions that hold for a mixed regular-chaotic phase space. These will be discussed in some detail in Section 3.2.1. Prior to evaluation of the distribution two properties need to be fulfilled. First, the levels considered must belong to a unique symmetry class (see the cases discussed in Section 1.2.2). Second, only the smooth part of the density of states must be considered and the mean level density must be normalised to unity as described in Section 3.2.1. Notice that the spectrum of the harmonic oscillator and also of the hydrogen atom are pathological cases, where no nearest-neighbour spacing distribution can be arranged reasonably.

Besides the famous nearest-neighbour spacing distribution RMT offers universal results for all measures of level correlation in chaotic spectra. Two-point correlations can be measured by the spectral rigidity Δ_3 or the number variance Σ^2 and also higher correlations can be addressed [79]. The distribution of the energy gaps of avoided crossings can be predicted by RMT [137]. For the quantum fidelity that we discuss in Chapter 4 RMT can characterise the decay behaviour in time of a chaotic system [53].

3.1 Classical considerations

For the consideration of quantum chaos in the following sections, it is necessary to give a characterisation of the classical phase space. The complete

3.1. Classical considerations

characterisation of the classical dynamics of a few-body system with chaotic behaviour is a challenging task on its own. Here we restrict ourselves to show that quasi-regular and chaotic trajectories coexist.

A characteristic property of a classically chaotic system is the sensitivity of trajectories to initial conditions. In our analysis of the phase space we aim to confirm this property in such a way, that we evaluate the dominant frequencies of trajectories with close lying initial conditions. In a regular system the dominant frequencies coincide with the fundamental frequencies in action-angle variables and depend smoothly on the initial conditions. For a mixed regular-chaotic system chaotic trajectories might stay close to regular tori for some time, perform chaotic motion and stick again to different regular tori. This is another consequence of the KAM theorem on dynamics in regular systems with small perturbations. The frequencies of trajectories in such systems reflect this behaviour and no smooth dependence on initial conditions can be observed, unless close to regular regions.

We consider trajectories in the time interval $[t_0, t_0 + T]$ with $t_0 = 100$ and $T = 400$. On each trajectory we perform the analysis of the dominant frequencies, as described in Section 2.4.2. Initial conditions are chosen such that the first electron starts on the x-axis with distance one half a.u. from the centre. For vanishing angular momentum $m = 0$ this electron starts at rest. For $m = 1$ it has momentum pointing in the y-direction $p_{y,1} = 2$, orthogonal to its position vector. The momentum of the first electron carries the total angular momentum. The second electron starts from equally distributed positions on the circle with radius one. Conveniently, we label its position in polar coordinates by the angle φ_2. For the second electron we let the momentum point outwards, parallel to its position vector. The absolute value of the momentum of the second electron adjusts the initial conditions to have the same energy. Typically we chose 1000 initial values on the half circle in different energy regimes. With increasing importance of the anharmonic potential for higher energies the trajectories show more and more chaotic properties. This is intuitively clear from the symmetry of the system and can be verified in terms of the analysis of the fundamental frequencies and their sensitivity to initial conditions. Typical scenarios for the frequencies are shown in Figure 3.1.

For $m = 0$ in the low-energy regime the regularity of the system is characterised by the smooth behaviour of the dominant frequencies with respect to the initial conditions (Figure 3.1 a)). In this case the COM is subject to approximately harmonic oscillations and the associated frequencies (squares and crosses) are constant and mostly degenerate. The frequencies of the relative

Chapter 3. Effects of the anharmonicity

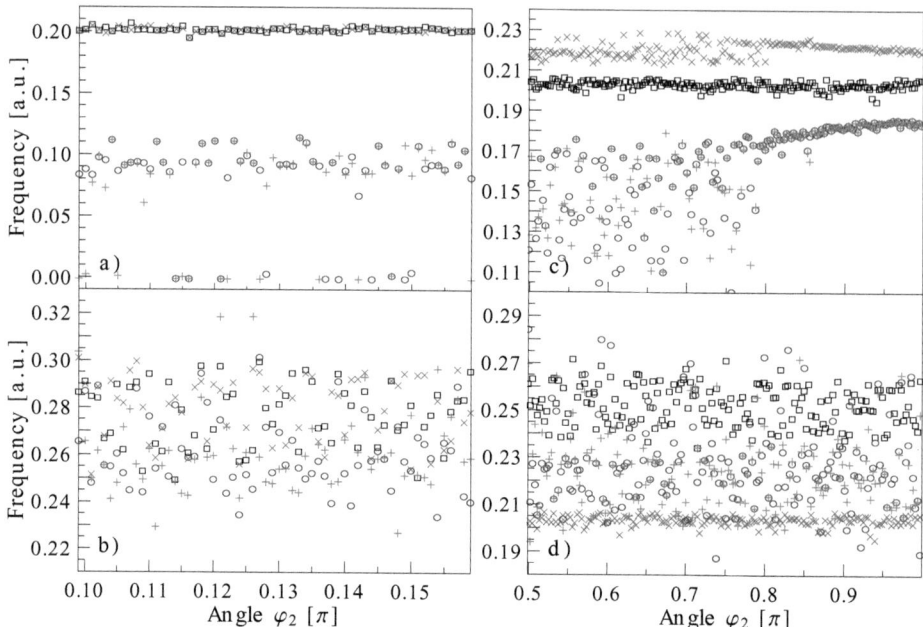

Figure 3.1: Dominant frequencies of the planar classical dot with $\omega_0 = 1$, $\gamma = 1$ and $\kappa = 0.1$. Depicted are the frequencies associated to the relative motion in x- (circles) and y- direction (pluses), and to the COM motion in x- (squares) and y- direction (crosses), respectively. For vanishing angular momentum a clearly regular motion is recognised in the low energy regime for $E = 3$ in a). For higher energy in b), $E = 15$, a dominantly chaotic dynamics is observed. The initial conditions are chosen, such that the first particle starts at rest from $x_1 = 0.5$ and $y_1 = 0$. The second electron starts from equally distributed positions on a circle of radius $r_2 = 1$ at the angle φ_2. The momentum of the second electron points outwards with varying absolute value to compensate the difference in Coulomb energy. For angular momentum $m = 1$, a mixed regular-chaotic dynamics is reflected in c) with $E = 5$. On the left-hand side of c) the frequencies are disordered, while on the right-hand side of c) the frequencies are nearly constant and partially degenerate. For $m = 1$ and $E = 10$ shown in d) only minor characteristics of regularity are dominated by generally disordered frequencies. The non-vanishing angular momentum is incorporated for the initial conditions by adding the appropriate momentum to the first electron in the y-direction.

motion (circles and pluses) change only little or can even be constant in some intervals. The dynamics is thus confined to regular nearly harmonic islands.

Features of a mixed phase space are observed for $m = 1$ and $E = 5$ in Figure 3.1 c). A regular, nearly harmonic island can be identified with the region, where $\varphi_2 \geq 0.8\,\pi$: All frequencies tend to be nearly constant and those associated to the relative motion are degenerate. For initial conditions with $\varphi_2 < 0.8\,\pi$ the COM motion appears to be nearly regular, while the frequencies of the relative motion are strongly disordered. This can be interpreted as a consequence of the chaotic motion.

An increase in energy leads to a generally discontinuous behaviour of the frequencies (see Figure 3.1 d)) and results in a complete lifting of the degeneracies observed above. While for all former cases some features of regularity remained, the frequencies for $E = 15$ in Figure 3.1 b) are totally disordered. Thus, the trajectories investigated in this regime are all subject to chaotic motion.

3.2 Measures of quantum complexity

Our analysis will focus on the dependence on the anharmonicity of $P(s)$, $P_{\mathrm{ac}}(c)$ and a measure of the localisation of eigenstates. We limit our calculations to zero angular momentum for all four symmetry classes (singlet/triplet spin symmetry and even/odd parity). The basis size was determined by the parameter $n_{\mathrm{base}} = 130$, which leads to basis sizes of approximately 12 000 basis vectors, where exact numbers depend on the symmetry class. Each calculation supplied more than 1000 well converged eigenvalues.

3.2.1 Spectral measures

In the case of mixed regular-chaotic dynamics, the nearest-neighbour spacing distribution $P(s)$ can, with some exceptions [62], be better described by a phenomenological distribution by Brody [60] than by a physically motivated distribution due to Berry and Robnik [61]. A remedy is found by extending the idea underlying the Berry-Robnik distribution, which is to split the classical phase space into distinct regular and chaotic regimes. The effects of dynamical tunnelling [141] and chaos assisted tunnelling [142] connect these two classically distinct regimes in the quantum regime and already improve the results [64, 63]. Recent additional achievements considering the effects of flooding [65] appear to complete the discussion, at least for the nearest-neighbour spacing distribution. Nevertheless a very thorough analysis of the classical phase space is necessary

Chapter 3. Effects of the anharmonicity

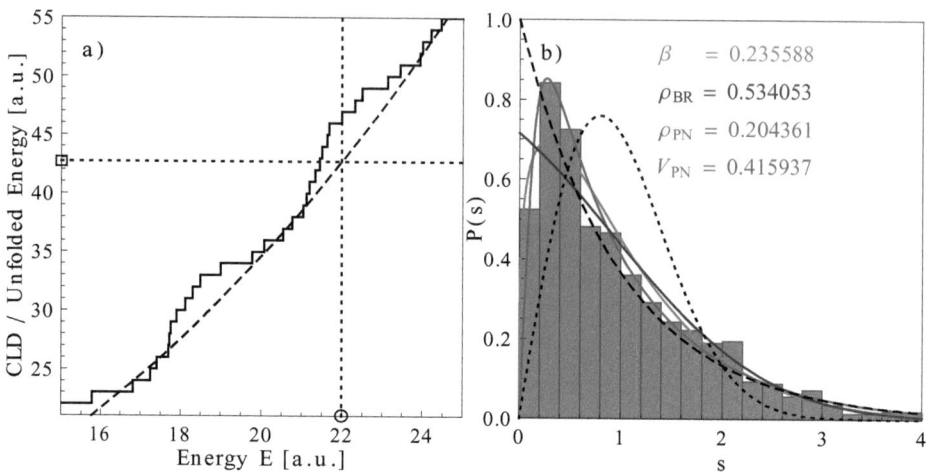

Figure 3.2: The procedure of unfolding and a typical result for the nearest-neighbour spacing distribution with appropriate fit functions. The procedure of unfolding is shown in a). The exact numerical eigenenergies are used to plot the cumulated level density (CLD) (solid line), which is interpolated by a polynomial fit (dashed line). A new unfolded energy ($E = 42.694687$, square) equals the value of the fitted CLD at the exact eigenenergy ($E = 21.990952$, circle) depicted by the dotted lines. The mean level spacing is thereby automatically normalised to unity. In b) the numerically obtained distribution of nearest-neighbour spacings after unfolding (bars) is fitted by the Brody distribution (green line), the Berry-Robnik distribution (blue line) and the Podolskiy-Narimanov distribution (red line). For reference we also show the limiting cases for regular dynamics, the Poisson distribution (dashed line), and for chaotic dynamics, the Wigner distribution (dotted line). The example data corresponds to singlet states with even parity, vanishing angular momentum, $\omega_0 = 1.0$, $\kappa = 0.1$ and $\gamma = 0.3$.

to obtain an *ab initio* description of this distribution. To our knowledge, this has only been performed for elementary systems.

The universal predictions of random matrix theory can only be confirmed if the characteristic spectral properties of a system are brought to some general footing. This procedure is called "unfolding" and different methods are in use. The common two steps for all methods are the following: (i) The level density is smoothed by a simple fit or semiclassical analysis and (ii) a new set of energies is derived from the smoothed level density in a way that the mean level density is normalised to unity. We fit the cumulated level density (CLD) obtained from the numerical exact eigenenergies E_n

$$n(E) = \sum_n \Theta(E - E_n), \qquad (3.3)$$

with $\Theta(x)$ the Heaviside step function, by a cubic polynomial (i) and take the value of the smoothened CLD evaluated at the former eigenenergy to be the new energy (ii). For clarity we demonstrate the procedure of unfolding in Figure 3.2 a). Alternative methods [137] produce qualitatively identical results.

The universality of chaotic properties in quantum mechanical systems has been shown for many examples involving the statistics of the separations of neighbouring energy levels. Once the limits for regular and purely chaotic dynamics were observed in different experimental and numerical studies, the interest in situations with a mixed classical phase space arose and tools to describe the smooth transition between these regimes were developed. If there are any symmetries left in the problem, the nearest-neighbour spacing distribution will most likely be a Poisson distribution. Otherwise, the level statistics of an underlying mixed phase space will exhibit a level repulsion which, however, is not as pronounced as in the pure chaotic case. Alternatives for modelling these statistics are the Brody distribution [60] characterised by the parameter β,

$$P_{\text{Brody}}(s) = (\beta + 1)as^\beta \exp(-as^{\beta+1}), \qquad (3.4)$$

with $a = \Gamma\left(\frac{\beta+2}{\beta+1}\right)^{\beta+1}$; the Berry-Robnik distribution [61] characterised by the parameter ρ_{BR},

$$P_{\text{BR}}(s) = \left[(1-\rho_{\text{BR}})^2 \operatorname{erfc}(\frac{\sqrt{\pi}}{2}\rho_{\text{BR}}s) + \left(2\rho_{\text{BR}}(1-\rho_{\text{BR}}) + \frac{\pi}{2}\rho_{\text{BR}}^3 s\right) e^{-\frac{\pi}{4}\rho_{\text{BR}}^2 s^2}\right] e^{-(1-\rho_{\text{BR}})s}; \qquad (3.5)$$

and the extension of the latter proposed by Podolskiy and Narimanov [63, 143] characterised by the parameters ρ_{PN} and V_{PN},

$$P_{\text{PN}}(s) = \left[(1-\rho_{\text{PN}})^2 F\left(\frac{s}{(V_{\text{PN}})^2}\right) \operatorname{erfc}(\frac{\sqrt{\pi}}{2}\rho_{\text{PN}}s) + \left(2\rho_{\text{PN}}(1-\rho_{\text{PN}})F\left(\frac{s}{V_{\text{PN}}}\right) + \frac{\pi}{2}\rho_{\text{PN}}^3 s\right) \times e^{-\frac{\pi}{4}\rho_{\text{PN}}^2 s^2}\right] e^{-(1-\rho_{\text{PN}})s}, \qquad (3.6)$$

with

$$F(x) = 1 - \frac{1-\sqrt{\frac{\pi}{2}x}}{e^x - x}.$$

The Brody parameter β describes the transition from regular ($\beta = 0$) to chaotic ($\beta = 1$) behaviour. It must be noted that the Brody parameter β lacks

Chapter 3. Effects of the anharmonicity

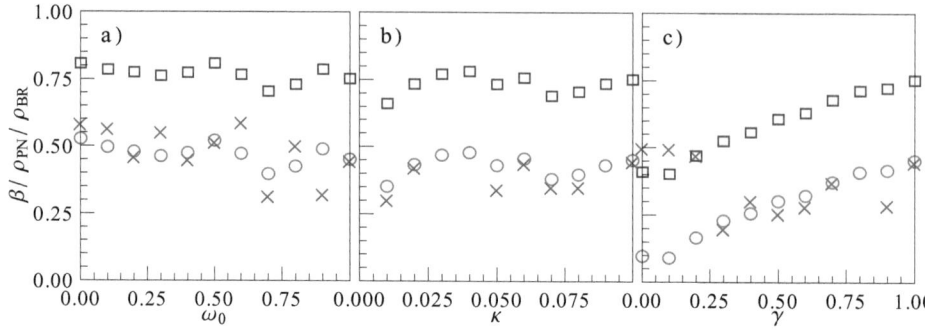

Figure 3.3: Fit parameters for the nearest-neighbour spacing distributions for variation of ω_0, κ and γ. In particular we show the fit parameters for the nearest-neighbour spacing distribution of the Brody distribution (β, green circles), the Berry-Robnik distribution (ρ_{BR}, blue squares) and the Podolskiy-Narimanov distribution (ρ_{PN}, red crosses). The states have spin singlet symmetry, even parity and vanishing angular momentum. In a) $\gamma = 1$ and $\kappa = 0.1$ are constant, while ω_0 is varied. In b) $\omega_0 = 1$ and $\gamma = 1$ are constant, while κ is varied. For $\kappa = 0$ the distribution is too close to the pathological harmonic oscillator case to give meaningful results. In c) $\omega_0 = 1$ and $\kappa = 0.1$ is constant, while γ is varied. For all three plots the furthermost right configurations coincide.

a quantitative physical meaning, but describes the transition merely qualitatively. The Berry-Robnik distribution is characterised by the parameter ρ_{BR}, which is the ratio of the chaotic to the total phase space volume and is a purely classical property of the system. The generalisation of the Berry-Robnik distribution by Podolskiy and Narimanov include the equivalent ratio ρ_{PN} and additionally perturbative quantum corrections. The parameter V_{PN} describes the tunnelling between regular and chaotic regions and chaos assisted tunnelling between regular regions via the chaotic sea [63].

We have considered all states with principal quantum numbers ranging from $n = 1$ to $n = 40$, in the unperturbed system, for all symmetries[1]. The grey bars in Figure 3.2 b) depict our numerical data for the specific case with $\omega_0 = 1$, $\kappa = 0.1$ and $\gamma = 0.3$. The different fits according to the previously described distributions are given by the coloured lines. For the Berry-Robnik distribution we fitted the parameter ρ_{BR}. The fit for the numerical data using the Brody distribution is in general better than the fit by the Berry-Robnik distribution, but has the disadvantage of an unphysical fit parameter. In the considered regime both distributions are similar for $s > 1$, which is the reason for the constant shift of approximately 0.3 between the parameters (see Figure 3.3).

---

[1]The specific number of eigenenergies shown in Table A.5.

While the Berry-Robnik distribution overestimates the number of very small spacings, the Brody distribution underestimates this number. The Podolskiy-Narimanov distribution improves the quality of the fits, naturally, as a second fit parameter V_{PN} is introduced.

We show the results of the statistical analysis for the singlet spin symmetry even parity case performed for 946 eigenvalues ($\alpha = 0.2, n_{\text{base}} = 130, n_{\text{tot}} = 11\,168$). The results for the different symmetry classes are qualitatively identical and we restrict our discussion to this exemplary case. We have used bins of width 0.2 for the histograms of the numerical nearest-neighbour spacing distribution. In order to get an overview of the parameter-dependence of the complexity we analysed the distributions for different configurations, see Figure 3.3. For the potential with $\gamma = 1$ and $\kappa = 0.1$ we varied the harmonic frequency $\omega_0 \in [0, 1]$. It was expected that the harmonic potential will not have a great influence on the complexity, since the separability is lost due to the interaction and the quartic potential. The results satisfy our expectation: in Figure 3.3 a) we can observe a nearly constant behaviour of the parameters. However, for $\omega_0 \geq 3$ the behaviour of the nearest-neighbour distribution is dominated by the harmonic confinement. Though we cannot describe this situation with the nearest-neighbour spacing distribution, it is obvious that the system becomes dominantly regular. For experimental realisations this is an interesting fact, since for a system that shows chaotic properties without magnetic field, already moderate field strengths introduce a great amount of regularity. This is expected to be true also for other (non-quartic) perturbations.

When varying the parameter κ for a quantum dot with interacting electrons ($\gamma = 1$) and harmonic confinement ($\omega_0 = 1$) we have to consider an exception. For $\kappa = 0$ the system equals Hooke's atom and we analysed the spectrum of the radial equation of the relative motion. Being still too close to the pathological harmonic oscillator case no proper fit could be performed. For small values of κ the harmonic confinement is still dominant. The transition to the complex regime happens in a rather small interval of $\kappa \in [4 \times 10^{-3}, 7 \times 10^{-3}]$. In Figure 3.3 b) we show the results for $\kappa \in [0.01, 0.1]$. The parameters exhibit a small increase for small κ values and are mainly constant after. The main feature introducing the complexity is the coupling of all four degrees of freedom induced by the loss of separability, here the separability in COM and relative coordinates.

We vary the interaction ($\gamma \in \{0.0, 0.1, .., 1.0\}$) with constant harmonic and quartic potential ($\omega_0 = 1$ and $\kappa = 0.1$). Thus, we start from two independent particles and arrive at the full potential model. For vanishing interaction the system is regular and the distribution is expected to be the Poisson distribution, which is reflected by our data. Nevertheless, the parameters in Figure 3.3 c) do not vanish for $\gamma = 0$, we will come back to this point in the next paragraph. With increasing interaction strength the parameters show a constant increase according to stronger mixing of the two independent motions.

There is some ambiguity involved about the evaluation of the nearest-neighbour spacing distribution and the different fits. The first ambiguity is the procedure of unfolding, but the results are robust for alternative fit functions. Indeed our cubic fit of the CLD is motivated by semiclassical analysis (the CLD of the harmonic oscillator for constant angular momentum is cubic). Secondly, the choice of the width of the bin for $P(s)$ introduces some arbitrariness. We analysed different widths and presented the data for the commonly used width of 0.2. When fitting the different fit functions of Brody, Berry and Robnik, and Podolskiy and Narimanov the influence of the widths is different. For P_{Brody} and P_{BR} the resulting parameters do not depend strongly on the width of the bins. Both fit distributions represent our numerical data for $s > 1$ very well. For small spacings $s < 1$, both distributions show deviations from the numerical data, caused by the quantum mechanical effects, mentioned above [64]. The distribution P_{PN} takes into account these effects and improves the fit for small spacings. For our system the two parameters of P_{PN} can hardly be derived and must be obtained by a fit to the numerical data. A variation of the widths of the bins leaves some arbitrariness to the fit parameters ρ_{PN} and V_{PN}. The resemblance of our system to a harmonic oscillator leads to pronounced peaks in the numerical distributions. For dominant harmonic confinement none of the fit functions is well suited. For most of the cases we find that the analysis of the nearest-neighbour spacing distribution offers a good qualitative characterisation of the complexity of the system.

3.2.2 Avoided crossings

A manifestation of the non-separability of a system and its complexity is the occurrence of avoided crossings in the spectrum depending on a slowly varying parameter. This is the case in our system when κ is the adiabatic parameter (see Figure 3.4 a)). Avoided crossings are naturally related to a drastic interchange of the symmetry properties of the eigenstates involved. This is typically characterised by abrupt changes in the behaviour of some expectation values. For instance, as the eigenstates $|n_c, m_c, n_r, m_r\rangle^{\epsilon_p} = |1, 4, 0, \text{-}4\rangle^+$ and $|0, 2, 3, \text{-}2\rangle^+$ of the harmonic ($\kappa = 0$) quantum dot *evolve* as κ varies, there is a sudden

3.2. Measures of quantum complexity

jump in the expectation value of the COM radial distance $\langle R_{COM} \rangle$ close to the avoided crossing around $\kappa = 0.00716$ as shown in Figure 3.4 c). Less pronounced is the change of $\langle R_{COM} \rangle$ for the states $|0,0,5,0\rangle^+$ and $|3,2,0,\text{-}2\rangle^+$ close to the avoided crossing around $\kappa = 0.00325$, though the respective wave functions completely interchange their properties as seen in the density plots on the right of Figure 3.4. The distribution $P_{\text{ac}}(c)$ is, in the mixed case, the sum of a δ-peak, representing real crossings in the regular regime, and a normal distribution for the avoided crossings in the chaotic regime [144]. For the construction of the distribution $P_{\text{ac}}(c)$ we require an efficient determination of avoided crossings in a large amount of spectral data. Though jumps in the expectation values, e.g. $\langle R_{COM} \rangle$, can be used to identify avoided crossings, the systematic detection of these jumps is not a trivial task. Alternatively, the *quantum fidelity susceptibility* of the eigenstates provides an efficient method for such purpose. The quantum fidelity susceptibility χ of an eigenstate ψ_n is equivalent to the curvature of the associated eigenenergy E_n depending on the varying parameter [145]. It can be calculated via the static quantum fidelity $F_{\delta\kappa}(\kappa, n) = |\langle \psi_{n,\kappa} | \psi_{n,\kappa+\delta\kappa} \rangle|^2$,

$$\chi = \lim_{\delta\kappa \to 0} \frac{1 - F_{\delta\kappa}}{(\delta\kappa)^2} \approx -\lim_{\delta\kappa \to 0} \frac{\log(F_{\delta\kappa})}{(\delta\kappa)^2}, \quad (3.7)$$

and is largely independent of the perturbation $\delta\kappa$ [146].

The typical behaviour of the quantum fidelity susceptibility close to an avoided crossing is illustrated in Figure 3.4 b). It is characterised by three properties, whose detection is easy to implement for practical purposes: The susceptibility has a peak near an avoided crossing, this peak is nearly identical for the two non-crossing states and the mean values of the susceptibilities before and after the peak interchange for these two states. The last property corresponds to the interchange of observables as seen for $\langle R_{COM} \rangle$ and can be observed for the avoided crossing at $\kappa = 0.00716$ in Figure 3.4 b).

For the detection of the avoided crossings we calculated the spectrum for singlet even parity states with $\omega_0 = 1$ and $\gamma = 1$ and vanishing angular momentum. The quartic potential was varied in two intervals $\kappa \in [0, 0.01]$ and $\kappa \in [0.04, 0.05]$ with a step size of $\Delta\kappa = 10^{-5}$ to improve the resolution of very narrow avoided crossings. After checking that the susceptibility is independent of the perturbation for several values of $\delta\kappa \in \{10^{-6}, 10^{-7}, 10^{-8}, 10^{-9}\}$ data collection was performed using $\delta\kappa = 10^{-7}$. With this method it was possible to determine over 13 000 avoided crossings within the first interval and 1378 eigenstates, equivalent to energies up to 70. In the second interval only 640 avoided

Chapter 3. Effects of the anharmonicity

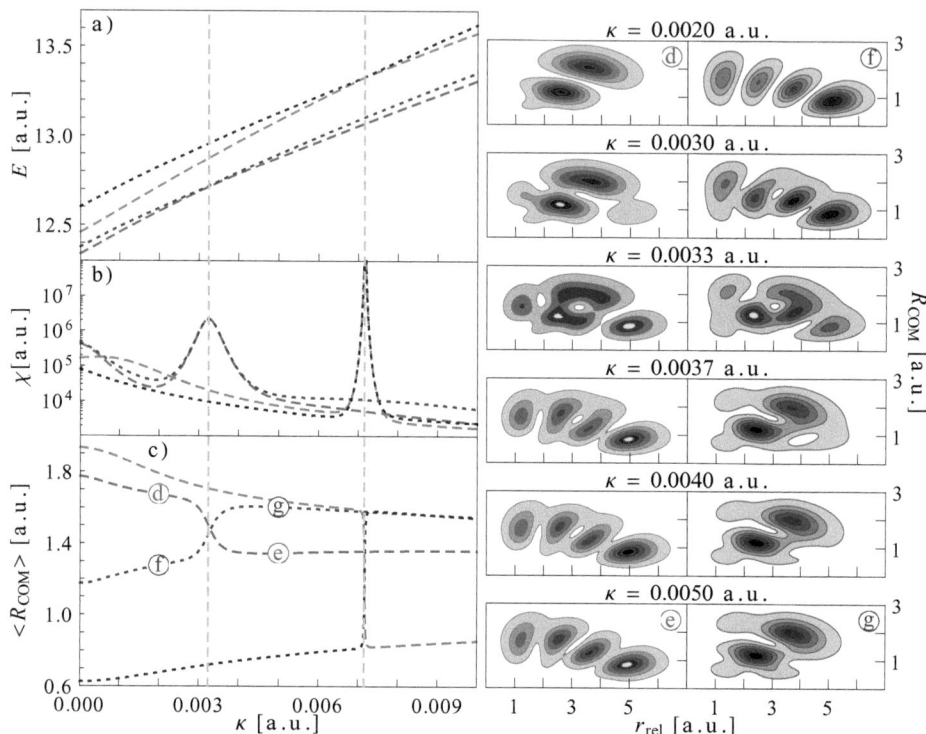

Figure 3.4: Characteristics and detection of two examples of avoided crossings occurring for variation of κ. The plot shows four singlet states with even parity and vanishing total angular momentum in the potential with $\omega_0 = 1.0$, $\gamma = 1.0$ and varying κ. The eigenstates can be labelled by the exact quantum numbers n_c, m_c, n_r and m_r of the harmonic problem ($\kappa = 0$). $|0,0,5,0\rangle^+$: red; $|3,2,0,-2\rangle^+$: blue; $|0,2,3,-2\rangle^+$: green; $|1,4,0,-4\rangle^+$: violet. In a) the avoided crossings can be observed in the spectrum. For the numerical detection of avoided crossings the quantum fidelity susceptibility shown in b) is a valuable tool. The susceptibilities of the two non-crossing states show peaks, which coincide at the maximum. The states interchange their properties during the avoided crossing. This is reflected by physical observables, e.g. the expectation value of the COM radial distance $\langle R_{\text{COM}} \rangle$ in c). It can also be visualised, even more impressively, by successive plots of the probability densities of the non-crossing states. This is shown on the right-hand side with contour plots of the states $|0,0,5,0\rangle^+$ and $|3,2,0,-2\rangle^+$ for different values of κ. The similarity of the interchanged states before and after the avoided crossings, d) and e), and f) and g), respectively, can be observed.

3.2. Measures of quantum complexity

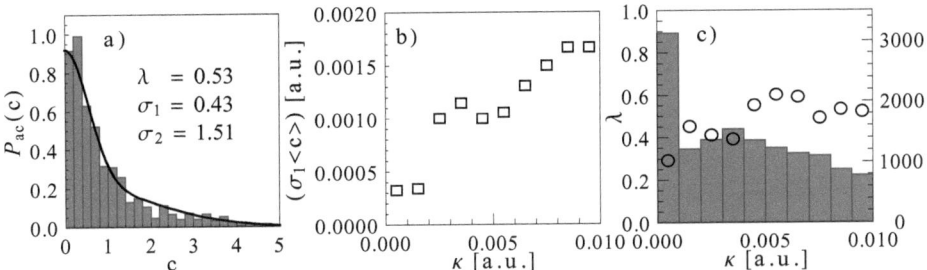

Figure 3.5: Characterisation of the distributions of the widths of the avoided crossings. In a) the numerical distribution of the widths of the avoided crossings in the sub-interval $\kappa \in [0.009, 0.01]$ for $\omega_0 = 1$ and $\gamma = 1$ (grey bars) is shown. The fit with the modified distribution (3.9) (solid line) gives the parameters λ, σ_1 and σ_2. The widths of the distributions of energy gaps of the avoided crossings for regular eigenstates $\sigma_1 \langle c \rangle$ is shown in b) (squares). The order of magnitude is in accordance with the estimated error for the numerically evaluated gaps due to the finite step width of $\Delta \kappa = 10^{-5}$. The estimate depends on the angle between the crossing solitonic states. This might explain the lower values in the first two sub-intervals. In c) we show the results for the fit parameter λ denoting the fraction of chaotic eigenstates (circles). The grey bars depict the number of avoided crossings found per sub-interval (see the associated scale on the right-hand side of the plot).

crossing could be detected for the same number of considered states.

For each avoided crossing the energy gap between the two close lying states, known as the width of the avoided crossing, is calculated. These widths are normalised to unity by division with the mean value of the widths $\langle c \rangle$. The expected distribution is the weighted sum of the distribution of the normalised widths of the avoided crossings for chaotic states with a δ-peak, which represents the non-avoided crossings of regular states [144]:

$$P_{\mathrm{ac}}(c) = (1 - \lambda)\delta(c) + \frac{2\lambda^2}{\pi} \exp\left(-\frac{\lambda^2 c^2}{\pi}\right), \qquad (3.8)$$

where λ is the fraction of chaotic eigenstates. Our first attempt was to obtain values for λ by fitting our data with the cumulated distribution, $\int_0^c P_{\mathrm{ac}}(c') dc'$, following Ref. [144]. Our numerical distributions do not show a pronounced peak for $c = 0$ (see Figure 3.5 a)) and the cumulated fit was rather poor in most cases. A remedy could be found by broadening the δ-peak to a Gaussian distribution consistent with a finite width of the measured energy gaps c. A simple analysis of the spectrum provides us with an estimate of the width of this distribution. The slope of the energy levels depending on κ is of the order of magnitude of $\Delta E / \Delta \kappa \approx 10^2$, which gives an error in the energy of $\Delta E \approx 10^{-3}$ considering our numerical stepwidth for κ, which is $\Delta \kappa = 10^{-5}$.

Chapter 3. Effects of the anharmonicity

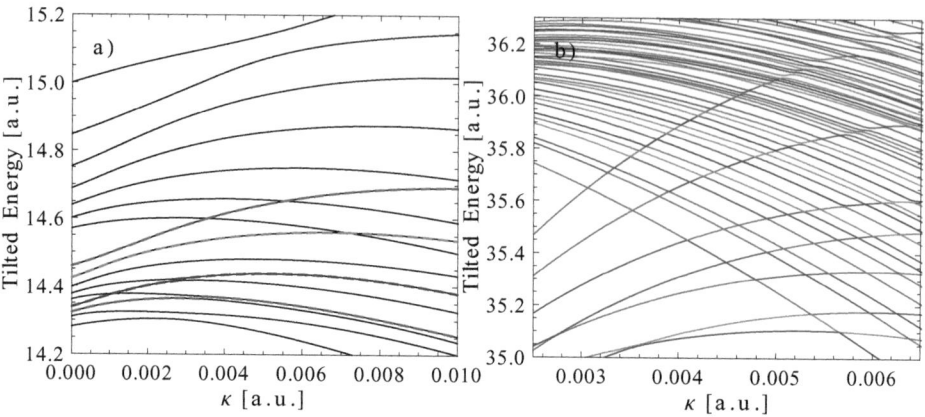

Figure 3.6: Characterisation of avoided level crossings between solitonic states. Level spectrum for varying κ of singlet spin states with even parity and vanishing angular momentum, $\omega_0 = 1$ and $\gamma = 1$. All observed level crossings are indeed avoided crossings, though in the majority of cases the energy gaps are too small to be observed in the plots. The values of the energies have been tilted, by substracting a linear fit of the mean energy slope, to obtain a better resolution of the level spacings. In a) all states emerging from the degenerate level $E = 14$ in the non-interacting harmonic limit are shown. As a guide to the eye we indicate the solitonic states $|1, 4, 1, -4\rangle^+$ (red), $|2, 4, 0, -4\rangle^+$ (blue), $|3, 2, 1, -2\rangle^+$ (green) and $|4, 2, 0, -2\rangle^+$ (magenta) by dashed lines. The slope of the tilt is $\Delta E/\Delta \kappa = 135$. In b) we show the two branches emerging from the degenerate levels with $E = 32$ and $E = 34$ in the non-interacting harmonic limit. The colours denote the different levels, the course of intersecting solitonic states is clearly recognisable. In the top right corner it can be observed how the solitonic states lose their character and randomly distributed larger energy gaps appear. The slope of the tilt is $\Delta E/\Delta \kappa = 700$.

We only ask for the total distribution to be normalised to unity and find the following distribution:

$$P_{\text{ac}}(c) = (1 - \lambda) \frac{2}{\pi \sigma_1} \exp\left(-\frac{c^2}{\pi \sigma_1^2}\right) + \lambda \frac{2}{\pi \sigma_2} \exp\left(-\frac{c^2}{\pi \sigma_2^2}\right), \tag{3.9}$$

with $\sigma_2 = \sigma_1 + (1 - \sigma_1)/\lambda$. The large amount of available data enables us to further divide the first interval in sub-intervals of width 0.001 and perform the analysis for states within these sub-intervals. With the fit of the distribution we determine the two independent fit parameters λ and σ_1 (see Figure 3.5 a) for the sub-interval $[0.009, 0.01]$). We find the real width $\sigma_1 \langle c \rangle$ to be of the order as expected from our simple analysis of the numerical error, see Figure 3.5 b). The quality of the fit could be confirmed by a χ^2-test of goodness of fit with a significance level of 1%.

Our results for the parameter λ are shown in Figure 3.5 c) denoted by circles, combined with the number of considered avoided crossings per sub-interval (grey bars). In the first sub-interval the number of avoided crossings is particularly high and the value of λ is the lowest found. In the following three sub-intervals λ stays below 0.5, while for $\kappa > 0.004$ the value of λ stays between 0.5 and 0.6, which is also true for the interval $[0.04, 0.05]$. Two features appear to be surprising: (i) the number of avoided crossings for $\kappa < 0.001$ exceeds the number of all following intervals $[k \times 10^{-3}, (k+1) \times 10^{-3}]$ for $k \in \{1,..,9\}$; (ii) the fraction of regular states that undergo avoided crossings is approximately 50% and even more for low values of κ.

We can give an interpretation of both effects by means of *solitonic* states, a concept introduced recently in [147]. The regular states responsible for the broadened δ-peak in the distribution P_{ac} correspond to solitonic states. The authors characterise these solitonic states by their smooth level dynamics with varying κ while they undergo avoided crossings, resembling the motion of solitonic waves. In our model we can identify two different cases, where the crossings of solitonic states appear. Levels with high angular excitation and high radial excitation in the COM motion tend to have a greater slope with varying κ than close-lying states, since they are more strongly affected by the increasing quartic potential. This leads to an intersection of levels within a branch of originally degenerate eigenstates in the non-interacting harmonic system (see Section 1.4.1). Indeed, both intersecting levels are considered as solitonic states, we show four examples in Figure 3.6 a) for the states $|1, 4, 1, -4\rangle^+$, $|2, 4, 0, -4\rangle^+$, $|3, 2, 1, -2\rangle^+$ and $|4, 2, 0, -2\rangle^+$.

This scenario occurs frequently in the sub-intervals with $\kappa < 0.003$ and is responsible for the high number of avoided crossings found in this regime, in particular in the first sub-interval. Notice that the choice of states in Figure 3.6 a) is for better visuability, since for $\kappa < 0.001$ we need to switch to higher energies, where very small level spacings and a by far greater number of involved levels occur.

The second case, where crossings of solitonic states occur in a great amount is at the intersection of two branches, see Figure 3.6 b). These branches, emerging from degenerate eigenenergies of the non-interacting harmonic model, start to intersect only for $\kappa > 0.001$ for the considered states. In the top right corner of Figure 3.6 b) the state starting from the energy $E \approx 35.48$ loses its solitonic character and the occurrence of truly chaotic avoided crossings with larger energy gaps can be observed.

Notice that the difference in the real widths of the regular states distribution in Figure 3.5 b) can be associated to the difference in the angle at which they intersect. The angles for crossings of solitonic states within a branch tend to be smaller than the angles for intersecting solitonic states from different branches.

3.2.3 Eigenvector localisation

We calculate an eigenstate specific information entropy depending on the parameters κ and γ, which depends on the basis representation of the eigenvectors. Following the arguments by Zelevinsky et al. [139] we obtain physically relevant results by choosing a reference basis for the representation of the eigenvectors that is physically related to the system under consideration. In our case this is the purely harmonic oscillator described in Section 1.2.2. A general eigenvector of this system is given by

$$|\psi_{nk}\rangle, \; n \in \{0,..,N\}, \; k \in \{1,..,K_n\},$$

where N goes to infinity and the values of K_n are given in Table A.5. We calculated numerically the harmonic case in order to achieve a complete (K_n) basis representation for all results presented here. Instead of determining the entropy directly in the harmonic oscillator basis, we project on the energy subspaces. For a general vector $|\varphi\rangle$, it holds

$$|\varphi\rangle = \sum_{n=0}^{\infty} \sum_{k=1}^{K_n} c_{nk} |\psi_{nk}\rangle = \sum_{n=0}^{\infty} e_n |E_n\rangle,$$

with

$$|E_n\rangle = \frac{1}{e_n} \sum_{k=1}^{K_n} c_{nk} |\psi_{nk}\rangle,$$

and the coefficients

$$e_n = \sqrt{\sum_{k=1}^{K_n} |c_{nk}|^2}$$

chosen such that the representation basis $|E_n\rangle$ is orthonormal.

The information entropy defined by

$$S_\varphi = \sum_{n=1}^{N} |e_n|^2 \log(|e_n|^2),$$

is a measure of the localisation of the vector $|\varphi\rangle$ in the harmonic oscillator basis. Large values of S_φ imply a large spread of the state $|\varphi\rangle$ in this basis. A similar analysis has been performed to identify the "solitonic" eigenstates in the spectrum of a tilted Bose-Hubbard model [147].

Figure 3.7 shows the entropies of eigenstates with singlet symmetry and even parity of a two-electron quantum dot with a harmonic confinement ($\omega_0 = 1$) and four different situations. As our reference basis is the harmonic oscillator basis the purely harmonic case gave only vanishing entropies and is not considered here. For the harmonic case with electron-electron repulsion, in a), a very regular behaviour of the entropies can be observed. Indeed, in this case the entropy is only a measure for the relative motion and the horizontally ordered entropies belong to states which differ only in the COM quantum numbers. The eigenstates of Hooke's atom presented in Table 2.1 are highlighted with circles and systematically belong to the states within the region of higher entropy. These states are configurations with no angular momenta in the subsystems of the COM and of the relative motion. The second class of states showing lower values for the entropy have non-vanishing angular momenta m_c and $m_r = -m_c$ in the subsystems. We have marked a series of these states with diamonds.

In Figure 3.7 b) we show the entropies with weak anharmonic confinement ($\kappa = 0.01$) and Coulomb repulsion. The behaviour of the lowest eigenstates does not change significantly, up to an energy of $E \approx 10$, while above this limit the entropies increase and lose most of their regular structure. The former effect can clearly be related to the narrowing of the potential and the higher number of harmonic states necessary to represent the eigenstates. The latter is a signature of the rising complexity of the system induced by the coupling of the two previously separate motions and the increasing number of avoided crossings. The occurrence of avoided crossings can in particular be seen from the jumps in the entropies of the states marked with circles, which we obtained by adiabatically following the levels from a) to b).

These effects are further enhanced for the stronger anharmonic ($\kappa = 0.1$) case in c), where only the ground-state entropy is close to the harmonic case. For the case in d) with anharmonic confinement but without electron-electron interaction the values of the entropies are as high as in c), but they show more specific structure. This can be understood by considering that, in this case, the system is again separable in two independent particles.

Qualitatively similar observations are obtained when we choose for instance the basis (2.32) as a reference.

Chapter 3. Effects of the anharmonicity

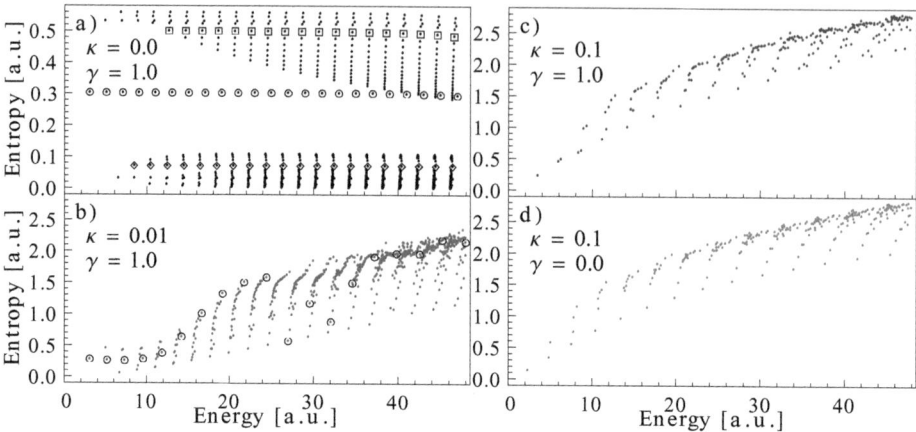

Figure 3.7: Information entropies of states with zero angular momentum, singlet symmetry and even parity for four combinations of the parameters κ and γ: a) $\kappa = 0.0$, $\gamma = 1.0$, b) $\kappa = 0.01$, $\gamma = 1.0$, c) $\kappa = 0.1$, $\gamma = 1.0$ and d) $\kappa = 0.1$, $\gamma = 0$. In all cases there is a harmonic confinement with $\omega_0 = 1$. The entropy has been calculated for approximately 1200 eigenstates in a), 700 eigenstates in b) and 250 eigenstates in c) and d). In a) we marked the states with the quantum numbers $|n_c, m_c, n_r, m_r\rangle^{\epsilon_p} = |k, 0, 0, 0\rangle^+$ with circles (also in b)), $|k, 0, 5, 0\rangle^+$ with squares and $|k, 2, 1, -2\rangle^+$ with diamonds, where k is an integer number. The other symmetry classes exhibit similar behaviour.

3.3 Summary

A brief introduction to the main ingredients of quantum chaos prepared the investigation of the complexity of our model system. The underlying classical phase space was demonstrated to be mixed regular-chaotic by means of sensitivity of dominant frequencies to initial conditions. In the spirit of the universal predictions of RMT we analysed the nearest-neighbour spacing distributions for all symmetry classes. The mixed phase space character and the vicinity to a harmonic oscillator left the applications of different fits of the distribution without quantitative success. Qualitatively the quantum spectrum was characterised to be on an intermediate chaotic level. A more profound characterisation could be given by the analysis of the avoided crossings for the system evolving under variation of the quartic potential strength. Quantum fidelity susceptibility proved to be a powerful tool to detect avoided crossings in large numerical spectra. The finite step-width for the numerical variation of κ was taken into consideration for the statistical analysis of the widths of the avoided crossings. In accordance with the former results the fraction of chaotic eigenstates indicates an intermediate level of chaoticity. The high number of avoided crossings, in particular for very small quartic potential strength, as well as the meaning of regular states undergoing avoided crossings could be interpreted by means of solitonic states. Finally a measure of the complexity extending the spectral measures to properties of the eigenstates was introduced. The localisation of an eigenstate in the harmonic oscillator reference basis was measured by the information entropy of the normalised vector representation. Small quartic potential strengths change the entropies of the regular case with purely harmonic confinement significantly. The localised states are spread over the reference basis states leading to higher values of the entropies and the highly ordered structure is lost. We could show that the full system exhibits mixed regular-chaotic properties in the classical and in the quantum regime. We will see further consequences in the following chapter considering time-dependent phenomena.

Chapter 4

Time dependence and quantum fidelity

Quantum mechanical time evolution in the Schrödinger picture is governed by the time dependent Schrödinger equation (TDSE),

$$i\partial_t |\psi(\boldsymbol{r},t)\rangle = \boldsymbol{H}(\boldsymbol{r},t)|\psi(\boldsymbol{r},t)\rangle. \tag{4.1}$$

It is a linear partial differential equation and exhibits the formal solution

$$|\psi(\boldsymbol{r},t)\rangle = e^{-i\boldsymbol{H}(\boldsymbol{r},t)t}|\psi(\boldsymbol{r},0)\rangle = \boldsymbol{U}(\boldsymbol{r},t)|\psi(\boldsymbol{r},0)\rangle, \tag{4.2}$$

with the unitary time evolution operator $\boldsymbol{U}(\boldsymbol{r},t)$[1]. In contrast to the classical, the quantum time evolution is unitary and insensitive to variation of initial conditions in terms of the wave function. The overlap of two arbitrary wave functions can be seen as an equivalent to a classical phase space distance. It is however constant in time in particular for two initial wave functions with slightly varying initial conditions.

In classical physics, in particular statistical mechanics, phenomena associated to chaos are very common. All classical systems can either be regarded as the limit of large actions of a microscopical quantum system or a quantum system under the influence of decoherence. Consequently, chaotic properties must inevitably be contained in the quantum dynamics, included in the static properties of the system in accordance with the results of the previous chapter. Nevertheless, a dynamical analysis of a system may be considered as a tool to reveal the chaotic properties [148]. The appropriate tool is the quantum Loschmidt echo, introduced by Jalabert and Pastawski [48] with this term, referring to the classical Loschmidt paradox of irreversibility in statistical mechanics. It describes the stability properties of the system under imperfect time reversal and was originally brought to discussion by Peres [52]. Instead of varying the initial conditions by perturbing the initial wave function, the Hamiltonian describing the time evolution is slightly perturbed. The Loschmidt echo

---
[1] Notice that bold symbols characterise general vectors, matrices or operators throughout this chapter.

is then defined by the overlap of the wave function evolved in time under the perturbed and unperturbed Hamiltonian, respectively. Addressing questions of quantum information theory the same measure is frequently used under the term quantum fidelity [149, 150]. In this community great interest in quantum fidelity arose as a model for the effect of decoherence on a small system (quantum bit) by interaction with the environment. The connection between quantum fidelity and measures of decoherence or equivalently the entanglement of a system with the environment has been established [49, 50, 51].

Predictions for the decay of the quantum fidelity can be given by RMT, semiclassical analysis or perturbation theory. Different decaying behaviours are well characterised, depending on the perturbation strength, the complexity and the dimension of the system, and also the initial state. A general review on quantum fidelity and its properties is given by Gorin *et al.* [53], while the review by Jacquod and Petitjean [151] emphasises the meaning of decoherence and irreversibility.

Most of the existing investigations of the Loschmidt echo are restricted to simple model systems, i.e. one-particle systems or spin chains, evolving under a given usually chaotic Hamiltonian. Investigations of the fidelity in many-body systems – which include trapped Bose-Einstein condensates [55], many electrons in non-parabolic quantum wells [56] and electron gases [54] – have shown an unusual behaviour of the quantum fidelity: it stays equal to unity until a critical time and then drops suddenly to much lower values. The common feature of these investigations is a mean-field approximation necessary for the description of such systems. A natural question that arises from this anomalous behaviour is whether this is a universal feature of many-body systems or a consequence of the mean-field treatment.

A similar effect has been observed by Prosen for regular [152] as well as for chaotic [153] dynamics and is denoted by quantum freezing. The perturbations that are responsible for the freezing of the quantum fidelity fulfil a certain property. All perturbations in a mean-field approximation might possess this property [53], thus relating the unusual behaviour in the mean-field treatment to quantum freezing.

Our model has all characteristics to emulate the calculation in non-parabolic quantum wells [56], but without mean-field approximation. In these investigations white noise was used as a perturbation, but the same effect was observed in the anharmonic quantum well using a quartic potential as a perturber [154]. For small perturbations the quantum fidelity remained constant for several hundreds of natural time units of the system. In order to calculate this effect in our model a very accurate method for the time evolution is indispensable. Considering the long evolution times it must additionally be a very fast method to make a detailed analysis of different parameters possible. We have put some effort into testing and development of sophisticated methods for the time evolution. We have successfully implemented different methods capable of computing the challenging task for long-time evolution of the quantum fidelity. These methods will be presented in some detail in Section 4.2. Preliminary results will be shown in Section 4.3.

4.1 Quantum fidelity decay

We briefly present the essential framework for quantum fidelity and use it as an indicator for the accuracy of different time-evolution methods in the following section.

4.1.1 Quantum fidelity

Quantum fidelity can most generally be defined for density matrices, including the dynamics of pure states, on which we will put our focus. For a general Hamiltonian with a small perturbation, $\boldsymbol{H}_\epsilon = \boldsymbol{H}_0 + \epsilon \boldsymbol{V}$, the quantum fidelity F_ϵ for a pure state $|\psi\rangle$ is defined by

$$F_\epsilon(t) = |\langle \psi_0(t) | \psi_\epsilon(t)\rangle|^2 = \left|\langle \psi | \boldsymbol{U}_0^\dagger(t)\boldsymbol{U}_\epsilon(t) | \psi\rangle\right|^2 = |\langle \psi | \boldsymbol{U}_0(-t)\boldsymbol{U}_\epsilon(t) | \psi\rangle|^2, \tag{4.3}$$

where ψ_ϵ and ψ_0 are the states evolving under the perturbed/non-perturbed time evolution operator ($\boldsymbol{U}_\epsilon(t)/\boldsymbol{U}_0(t)$), respectively. The quantum fidelity can be regarded as the squared modulus expectation value of the echo operator

$$\boldsymbol{M}_\epsilon(t) = \boldsymbol{U}_0(-t)\boldsymbol{U}_\epsilon(t). \tag{4.4}$$

The echo operator can be considered as the propagator in the interaction picture due to the relation

$$\frac{\mathrm{d}}{\mathrm{d}t}\boldsymbol{M}_\epsilon(t) = -\mathrm{i}\epsilon\tilde{\boldsymbol{V}}(t)\boldsymbol{M}_\epsilon(t), \tag{4.5}$$

with the perturbation in the interaction picture $\tilde{\boldsymbol{V}}(t) = \boldsymbol{U}_0(-t)\boldsymbol{V}\boldsymbol{U}_0(t)$. The formal solution for the propagator can be expanded in orders of ϵ to give the linear response description for high fidelity ($F_\epsilon(t) \approx 1$):

$$F_\epsilon(t) = 1 - \epsilon^2 \int_0^t \mathrm{d}t' \int_0^t \mathrm{d}t'' \left[\langle\psi|\tilde{\boldsymbol{V}}(t')\tilde{\boldsymbol{V}}(t'')|\psi\rangle - \langle\psi|\tilde{\boldsymbol{V}}(t')|\psi\rangle\langle\psi|\tilde{\boldsymbol{V}}(t'')|\psi\rangle\right] + \mathcal{O}(\epsilon^4). \tag{4.6}$$

Chapter 4. Time dependence and quantum fidelity

The first non-vanishing term in this expansion is called the quantum fidelity susceptibility $\chi(t)$, just like in the static case (3.7). It corresponds to the linear response of the fidelity to the perturbation.

The characteristic property that leads to the effect of freezing is the vanishing of the diagonal elements of the perturbation operator, $\bar{V} = \lim_{t\to\infty} \frac{1}{t}\int_0^t \tilde{V}(t')\mathrm{d}t'$.

4.1.2 Quantum fidelity for eigenstates

Eigenstates of the unperturbed system, $H_0|N\rangle = E_N|N\rangle$, build a special class of initial states for the quantum fidelity. It reduces to the following expression

$$F_\epsilon(t) = |\langle N|U_0(-t)U_\epsilon(t)|N\rangle|^2 = \left|\langle N|\mathrm{e}^{-\mathrm{i}H_\epsilon t}|N\rangle\right|^2, \qquad (4.7)$$

describing a kind of "survival probability" for the initial state evolved under the perturbed Hamiltonian. In the linear response approximation we derive the following expression for the susceptibility,

$$\chi(t) = 2\sum_{k\neq N} \frac{|\langle N|V|k\rangle|^2}{(E_k - E_N)^2}\left(1 - \cos((E_k - E_N)t)\right), \qquad (4.8)$$

depending only on static properties of the unperturbed system. This is in accordance with the result that the survival probability equals the Fourier transform of the local density of states with respect to the eigenstate [155].

4.2 Numerical methods for the time evolution

The TDSE in matrix representation can be reduced to the standard initial value problem also in our more general framework with non-trivial overlap matrix

$$\partial_t \boldsymbol{y} = \boldsymbol{f}(\boldsymbol{y}, t), \ \boldsymbol{y}(t=0) = \boldsymbol{y}_0, \qquad (4.9)$$

where $\boldsymbol{f}$ is a d-dimensional linear function. A variety of mathematical tools for solving this general class of ordinary differential equations is available. A wide energy range of the spectrum creates a problem, usually referred to as *stiff*, most easily characterised by the ratio between highest and lowest eigenvalue of the evolution matrix ($\boldsymbol{f}$). This kind of problem leads to divergences for simple methods, like the Euler methods or the well-known explicit Runge-Kutta method. A reduction of the time step can partially avoid the divergences, but is not a feasible way for long-time evolutions. One way to overcome these problems is to use implicit methods, which are always computationally very demanding, since at each time step at least one matrix inversion has to be calculated. Another way is to use explicit methods that are particularly suited to address stiff initial value problems.

The stiffness of our model and the high accuracy needed for calculating the quantum fidelity for long times and very small perturbations require the application of sophisticated time evolution methods. We test two explicit methods, one is based on an idea of Fatunla [156, 157] and the other one is a Krylov subspace method proposed by Arnoldi [158]. The former method takes into account the frequencies involved and has been shown capable of calculating the time evolution with large, stiff matrices [120]. Furthermore the error of the method is easily available at each step and can be used to adapt the time step efficiently. The latter method is a unitary method that overcomes the stiffness by reducing the effective spectrum of the problem while offering high accuracy and very low computation times [159]. A drawback of this method is that the local error is not available and therefore there is no canonical way to adapt the time step. Both explicit methods are included in an iterative implicit predictor-corrector scheme that has further been developed in the course of this work. A preservation of the norm up to machine precision could be achieved, even for highly demanding problems. As a new feature we included an adaption of the time step based on the error in the norm measured in units of the machine precision. However, the calculation times with this method are significantly increased in comparison with the two explicit methods. The method is particularly capable of propagating time-dependent Hamiltonians with a very high accuracy and will unfold its entire effectiveness in future applications.

To facilitate comparison all results presented in this chapter have been calculated on a single core of a Intel® Core™ 2 Quad CPU Q9400 (2.66GHz) with 8GB main memory.

4.2.1 Solving the time dependent Schrödinger equation

In our framework of a GEVP the TDSE (4.1) can be written as

$$\mathrm{i}\,\boldsymbol{B}\,\partial_t|\psi(\boldsymbol{r},t)\rangle = \boldsymbol{A}\,|\psi(\boldsymbol{r},t)\rangle. \tag{4.10}$$

We can now reduce to the standard initial value problem. Since $\boldsymbol{B}$ is positive-definite we can perform a Cholesky decomposition and obtain a lower triangular matrix $\boldsymbol{B} = \boldsymbol{L}\boldsymbol{L}^T$. The multiplication with the inverse triangular matrix is easily performed by backward substitution and we obtain the linear initial value problem we wish to solve:

$$\partial_t \boldsymbol{y} = \partial_t \left(\boldsymbol{L}^T |\psi(\boldsymbol{r},t)\rangle\right) = -i\left(\boldsymbol{L}^{-1}\boldsymbol{A}\boldsymbol{L}^{-T}\right)\left(\boldsymbol{L}^T|\psi(\boldsymbol{r},t)\rangle\right) = \boldsymbol{f}(\boldsymbol{y},t). \quad (4.11)$$

Note, that the evolved vector $\boldsymbol{y} = \boldsymbol{L}^T|\psi(\boldsymbol{r},t)\rangle$ now has trivial metric, since $\boldsymbol{y}^\dagger \boldsymbol{y} = \langle\psi|\boldsymbol{L}\boldsymbol{L}^T|\psi\rangle$. With the matrix representing the Hamiltonian dynamics of the system, $\boldsymbol{H} = \boldsymbol{L}^{-1}\boldsymbol{A}\boldsymbol{L}^{-T}$ we can write the initial value problem in the linear matrix representation

$$\partial_t \boldsymbol{y} = -i\boldsymbol{H}\boldsymbol{y}, \ \boldsymbol{y}(t=0) = \boldsymbol{y}_0. \quad (4.12)$$

4.2.2 Fatunla method

The described method goes back to work done by Fatunla over 30 years ago and takes into account the natural frequencies of the system [156, 157]. Its application to stiff systems has been tested and it has been applied successfully to the single ionisation of atoms by strong fields [120]. An exceptional feature of this method is the inclusion of the time derivatives of the evolution function $\boldsymbol{f}$ (see Equation (4.9)) that makes it particularly suited for time-dependent problems. Still, the way the intrinsic frequencies are included in the time evolution makes its application equally interesting for time-independent stiff problems like the one discussed here. Furthermore, the error in each step can be calculated explicitly and the time step δt can be adapted in a straightforward manner. This is also an advantage for time-independent problems, since the method finds an optimised time step automatically.

A recursive formula for the iteration of the vector $\boldsymbol{y}_n$ at time t_n along the time step δt is derived from an ansatz for the solution $\boldsymbol{y}(t)$. For a stiff initial value problem the solution can be approximated by an oscillating function

$$\boldsymbol{F}(t) = (\mathbb{1} - e^{\boldsymbol{\Omega}_1 t})\boldsymbol{a} - (\mathbb{1} - e^{-\boldsymbol{\Omega}_2 t})\boldsymbol{b} + \boldsymbol{c}, \quad (4.13)$$

with the diagonal stiffness matrices $\boldsymbol{\Omega}_i = \text{diag}(\Omega_{i,1},..,\Omega_{i,d})$ for $i \in \{1,2\}$ and constant vectors $\boldsymbol{a}$, $\boldsymbol{b}$ and $\boldsymbol{c}$. Assuming that $\boldsymbol{F}(t_n)$, $\boldsymbol{F}'(t_n)$ and $\boldsymbol{F}''(t_n)$ coincide with $\boldsymbol{y}_n$, $\boldsymbol{f}(t_n, \boldsymbol{y}_n)$ and $\boldsymbol{f}'(t_n, \boldsymbol{y}_n)$, respectively, $\boldsymbol{y}_{n+1}$ can be derived from $\boldsymbol{F}(t)$ at $t_{n+1} = t_n + \delta t$:

$$\boldsymbol{y}_{n+1} = \boldsymbol{y}_n + \boldsymbol{R}\boldsymbol{f}_n + \boldsymbol{S}\boldsymbol{f}_n^{(1)}, \quad (4.14)$$

where we use the notation $\boldsymbol{f}_n = \boldsymbol{f}(t_n, \boldsymbol{y}_n)$, $\boldsymbol{f}_n^{(k)} = \left.\frac{d^k}{dt^k}\boldsymbol{f}(t,\boldsymbol{y}_n)\right|_{t=t_n}$, and $\boldsymbol{R}$ and $\boldsymbol{S}$ are diagonal matrices defined by

$$\boldsymbol{R} = \boldsymbol{\Omega}_2\boldsymbol{\Phi} - \boldsymbol{\Omega}_1\boldsymbol{\Xi}, \qquad \boldsymbol{S} = \boldsymbol{\Phi} + \boldsymbol{\Xi}, \quad (4.15)$$

4.2. Numerical methods for the time evolution

with the diagonal matrices $\mathbf{\Phi}$ and $\mathbf{\Xi}$

$$\Phi_j = \frac{e^{\Omega_{1,j}\delta t} - 1}{\Omega_{1,j}(\Omega_{1,j} + \Omega_{2,j})}, \qquad (4.16)$$

and

$$\Xi_j = \frac{e^{-\Omega_{2,j}\delta t} - 1}{\Omega_{2,j}(\Omega_{1,j} + \Omega_{2,j})}. \qquad (4.17)$$

The stiffness matrices $\mathbf{\Omega}_1$ and $\mathbf{\Omega}_2$ can be written as

$$\begin{aligned}\Omega_{1,j} &= \frac{1}{2}\left(-D_j + \sqrt{D_j^2 + 4E_j}\right), \\ \Omega_{2,j} &= \Omega_{1,j} + D_j,\end{aligned} \qquad (4.18)$$

where D_j and E_j, for $j = 1, ..., d$ are given in terms of the components of the function $\mathbf{f}_n$ and its derivatives up to third order in t,

$$D_j = \frac{f_{n,j}^{(0)} f_{n,j}^{(3)} - f_{n,j}^{(1)} f_{n,j}^{(2)}}{f_{n,j}^{(1)} f_{n,j}^{(1)} - f_{n,j}^{(0)} f_{n,j}^{(2)}}, \qquad E_j = \frac{f_{n,j}^{(1)} f_{n,j}^{(3)} - f_{n,j}^{(2)} f_{n,j}^{(2)}}{f_{n,j}^{(1)} f_{n,j}^{(1)} - f_{n,j}^{(0)} f_{n,j}^{(2)}}. \qquad (4.19)$$

The local truncation error at $t = t_{n+1}$ is the difference between the exact solution at t_{n+1} and the numerical solution obtained. The analytical expression can be derived from the Taylor expansion in δt around $t = t_n$:

$$\begin{aligned}T_{n+1,j} = \frac{\delta t^5}{5!}\Big[&f_{n,j}^{(4)} + \left(\Omega_{2,j}^3 - \Omega_{2,j}^2\Omega_{1,j} + \Omega_{2,j}\Omega_{1,j}^2 - \Omega_{1,j}^3\right)f_{n,j}^{(1)} \\ &-\Omega_{1,j}\Omega_{2,j}\left(\Omega_{1,j}^2 - \Omega_{1,j}\Omega_{2,j} + \Omega_{2,j}^2\right)f_{n,j}^0\Big] + \mathcal{O}(\delta t^6).\end{aligned} \qquad (4.20)$$

By imposing a boundary criterion for $|\mathbf{T}_{n+1}|$ we can adapt the time step.

We try different limits for the truncation error $|\mathbf{T}_{n+1}|$ in order to test the performance of the method. We use the separable Hooke's atom with $\omega_0 = 1$ and $\gamma = 1$ for $\mathbf{H}_0$ and a quartic perturbation $\mathbf{V} = r_1^4 + r_2^4$ with the perturbation strength $(\epsilon)\,\delta\kappa = 10^{-5}$. The initial eigenstate is the state $|2,0,0,0\rangle^+$ with singlet spin symmetry, even parity, vanishing angular momentum and eigenenergy $E = 7$. We compute the quantum fidelity for eigenstates in a basis of 2370 vectors. The relative error of the susceptibility is the absolute value of the difference $\Delta\chi$ between the analytical expression χ in (4.8) and the result from the time evolution, divided by χ. As it is expected for a stiff problem the time step decreases to the lower bound of 10^{-5} during the calculation (see Figure 4.1 c)). Furthermore, we find the error in the norm to be approximately 10^{-9}, depending on the limits of the truncation error, which is a very accurate result

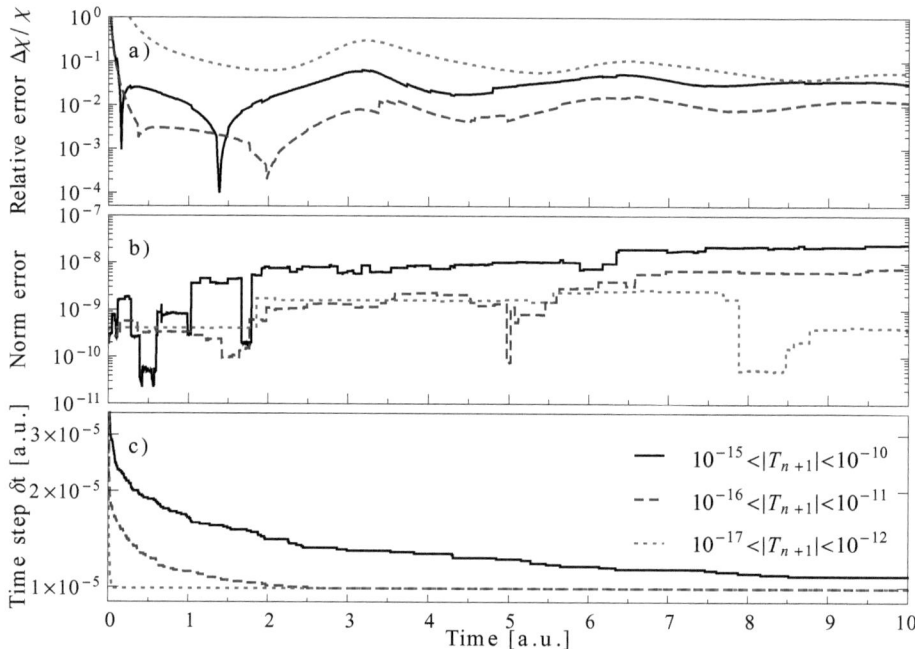

Figure 4.1: Properties of the Fatunla time evolution method calculating the quantum fidelity susceptibility of an eigenstate. For Hooke's atom we perturb the system via the quartic confinement $\delta\kappa = 10^{-5}$. The initial vector is the eigenstate $|2,0,0,0\rangle^+$ with zero angular momentum, singlet spin, even parity and eigenenergy $E = 7$. We use different bounds for the truncation error for the Fatunla method, as indicated in the lower right corner. The total computation time depends on these bounds: 5 days, 19 hours (black solid line); 7 days, 4 hours (blue dashed line); 7 days, 9 hours (red dotted line). In a) the relative error of the quantum fidelity susceptibility $\Delta\chi/\chi$ is shown. It is the absolute value of the difference $\Delta\chi$ between the analytical expression χ in (4.8) and the result from the time evolution, divided by χ. It is lowest ($< 10^{-2}$) for the intermediate bounds of the truncation error. In contrast, the error in the norm in b) is most of the time lowest for the bounds $10^{-17} < |T_{n+1}| < 10^{-12}$. We have set a lower bound of 10^{-5} for the time step. The evolution of the time step is depicted in c); as a consequence of the stiffness the time step drops to its lower bound after some time.

for a method that is not unitary (see Figure 4.1 b)). The relative error of the susceptibility is acceptable, though it might fall off in quality for longer time evolutions (see Figure 4.1 a)). The total computation time is approximately one week for the time range presented.

The Fatunla method is well-suited for time-dependent problems, in particular, when different regimes make an adaption of the time step strongly necessary. For our purpose of long time evolutions of a time-independent problem the Fatunla method shows very good results with minor errors for the norm and the quantum fidelity susceptibility. We expect to improve our results and in particular reduce the computation time by application of a unitary method.

4.2.3 Arnoldi method

The second explicit method we consider is a Krylov subspace method, just like the diagonalisation algorithm explained in Section 2.2.1. The technique was originally proposed by Arnoldi for partial diagonalisation of large matrices [158]. We briefly recall the time evolution method based on this diagonalisation technique that has been presented and successfully applied to laser-driven atomic helium in [159]. Since then a variety of time-dependent problems has been addressed with this method, e.g. [160, 161]. The central idea is to diagonalise the evolution operator in the Krylov subspace where a unitary time evolution is performed. A typical dimension of the Krylov subspace lies between 5 and 20, even for large basis sizes. The accuracy of the time evolution is strongly dependent on the Krylov dimension. From experience the smallest dimension possible gives the best results, where possible means, that a smaller dimension leads to a sudden breakdown of the convergence of the time evolution.

Assuming that the time interval is sufficiently small the Hamiltonian $\boldsymbol{H}$ may be treated as constant in time over a time step δt. We can approximate the time evolution during one time step according to Equation (4.12) by

$$\boldsymbol{y}(t+\delta t) = e^{-\mathrm{i}\boldsymbol{H}(t)\delta t}\boldsymbol{y}(t). \tag{4.21}$$

Equivalently we can express the time evolution by the series expansion of the exponential function

$$\boldsymbol{y}(t+\delta t) = \left(\mathbb{1} - \mathrm{i}\delta t \boldsymbol{H}(t) + \cdots + \frac{(-\mathrm{i}\delta t)^k}{k!}\boldsymbol{H}^k(t) + \cdots\right)\boldsymbol{y}(t). \tag{4.22}$$

Inspired by this representation we define the m-dimensional Krylov subspace K_m as

$$K_m = \mathrm{span}\{\boldsymbol{y}, \boldsymbol{H}\boldsymbol{y}, \cdots, \boldsymbol{H}^{m-1}\boldsymbol{y}\}, \tag{4.23}$$

and find an orthonormal basis set $\{q_0, q_1, ..., q_{m-1}\}$ by standard Gram-Schmidt orthogonalisation of the vectors in K_m, starting with $q_0 = y/|y|$. If we define Q to be a matrix formed by the m column vectors $(q_0, ..., q_m)$ the Hamiltonian can be projected on the Krylov subspace by

$$HQ = Qh \quad \Rightarrow \quad h = Q^\dagger H Q, \qquad (4.24)$$

with the m-dimensional quadratic Hamiltonian matrix h. The diagonalisation of h,

$$h = E \Lambda E^\dagger, \qquad (4.25)$$

with a diagonal matrix Λ and a unitary matrix E, can be performed straightforwardly with standard techniques, since it is in general a low-dimensional problem. Furthermore, properties of the full representation H inherited by the matrix h simplify the diagonalisation.

There are two major improvements in this method compared to the Fatunla method. First, the computation is particularly fast, since part of the calculation is performed in the reduced dimension of the Krylov subspace. Second, the norm is preserved with a very high accuracy. The only disadvantage is that no local truncation can be estimated and as a consequence it is not clear how to adapt the time step. The optimal dimension of the Krylov subspace is not known *a priori* and must be tested in combination with the time step. An optimal pair of Krylov dimension and time step can reduce the computation time and increase the accuracy at the same time.

As an example we calculated the quantum fidelity susceptibility for the same parameters as for the Fatunla method ($\omega = 1$, $\kappa = 0$, $\gamma = 1$, $\delta\kappa = 10^{-5}$, $m = 0$, $\epsilon_s = 1$, $\epsilon_p = 1$, $|2, 0, 0, 0\rangle^+$ and $n_{tot} = 2370$). We vary the Krylov dimension and the time step for several values and show results for three combinations in Figure 4.2. The computation times for the three examples are significantly different. The computation with the Krylov dimension of 5 and the time step of 10^{-4} (solid line) took approximately 4 hours, with Krylov dimension 7 and the time step 5×10^{-4} (dashed line) it took 45 minutes, while with Krylov dimension 9 and time step 10^{-3} it took 30 minutes. In all three cases the relative error of the susceptibility is better than with the Fatunla method, while all calculations are significantly faster.

The method of Arnoldi appears to be well suited for our system. The quantum fidelity in general and in particular for eigenstates can be interpreted as a measure for the deviation of the norm of the state from unity. Therefore, the conservation of the norm that is given up to at least 10^{-11} during the calculation is a perfect condition for our purpose. Since our Hamiltonian is not

4.2. Numerical methods for the time evolution

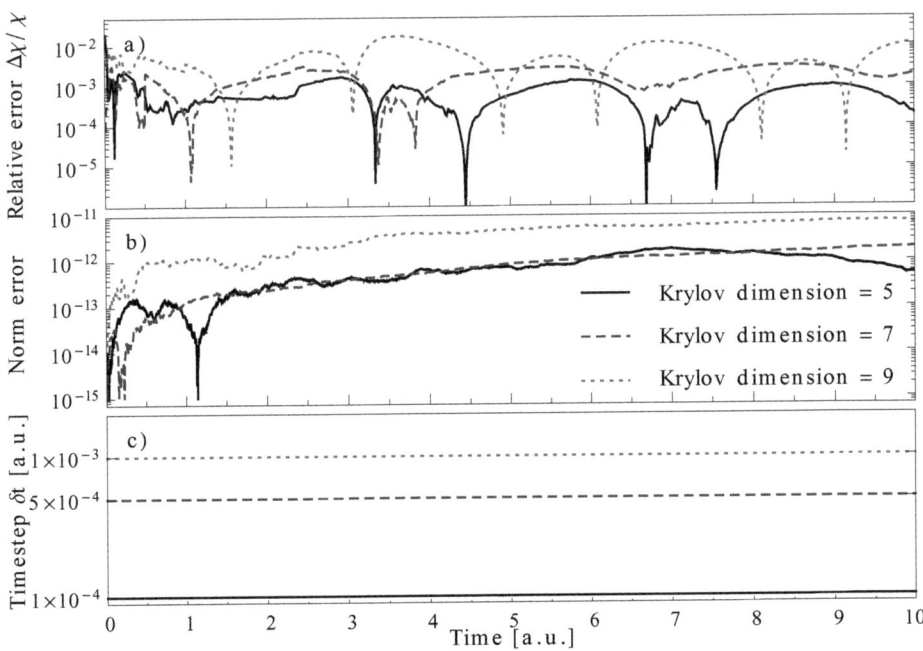

Figure 4.2: Properties of the Arnoldi time evolution method calculating the quantum fidelity susceptibility of an eigenstate. The configuration coincides with Figure 4.1 except for the time evolution method. We show three results with different dimensions of the Krylov subspace and different time steps. The relative error of the susceptibility shown in a) is in average one order of magnitude smaller compared to the Fatunla method. The error in the norm in b) is at least two orders of magnitude smaller than for the Fatunla method. Together with the significantly reduced calculation times of at most six hours the performance of the Arnoldi method is by far better than the Fatunla method. In c) the constant time step is shown. Decreasing the dimension or increasing the time step leads to divergent results for the cases presented here.

Chapter 4. Time dependence and quantum fidelity

time-dependent, the constant time step is no disadvantage, as soon as the optimal value is found. With little effort the time evolution can be carried forward to times of 100 or more without a significant loss of accuracy and within feasible computation times.

4.2.4 Advances for an iterative predictor-corrector scheme

Finally, we also consider an iterative predictor-corrector scheme, that has been presented by Hamido et al. [162] based on earlier works [163, 164]. General predictor-corrector methods use a simple time evolution method to predict the iterated vector y_{n+1} from y_n. In a second step a more sophisticated method, for example a high order implicit Runge-Kutta method, is used to correct the error made by the predictor.

In this approach the original time step δt is split into four intermediate time steps τ_i with $i \in \{1..4\}$ according to the four-step corrector method. The predictor, one of the two introduced explicit methods, is used to calculate the intermediate vectors $Y_i^{(0)}$ at the corresponding time steps τ_i. The corrector is based on an implicit Radau-type four-step Runge-Kutta method (see Appendix A.2). To correct the approximate solutions of the predictor this method is modified to give an iterative method with the four predicted solutions $Y_i^{(0)}$ as initial vectors,

$$\begin{pmatrix} Y_1^{(j)} \\ \vdots \\ Y_4^{(j)} \end{pmatrix} + i\delta t \begin{pmatrix} d_{11} H(t_n + \tau_1) & & 0 \\ & \ddots & \\ 0 & & d_{44} H(t_n + \tau_4) \end{pmatrix} \begin{pmatrix} Y_1^{(j)} \\ \vdots \\ Y_4^{(j)} \end{pmatrix} =$$

$$\begin{pmatrix} y_n \\ \vdots \\ y_n \end{pmatrix} - i\delta t \begin{pmatrix} (a_{11} - d_{11}) H(t_n + \tau_1) & \cdots & a_{14} H(t_n + \tau_4) \\ \vdots & \ddots & \vdots \\ a_{41} H(t_n + \tau_1) & \cdots & (a_{44} - d_{44}) H(t_n + \tau_4) \end{pmatrix} \begin{pmatrix} Y_1^{(j-1)} \\ \vdots \\ Y_4^{(j-1)} \end{pmatrix}$$

where $a = (a_{ij})$ and $b = (b_i)$ are the parameters of the Runge-Kutta-method (see Appendix A.2) and

$$d = \mathrm{diag}(d_{11}, .., d_{44})$$

is an appropriately chosen diagonal matrix to ensure fast convergence (see Appendix A.2). At each iteration step four linear equations of the full dimension d need to be solved. These linear equations are solved with the iterative bi-conjugate gradient stabilised method (Bi-CGSTAB) [165].

The predictor-corrector scheme can preserve the norm up to machine precision (10^{-15}), but is computationally very demanding. It is particularly suited for problems that need a very high accuracy. A possibility to adapt the time step reduces the computation time, in particular for systems with different regimes of the time evolution. This frequently occurs when the interaction of a laser pulse with an atom or equivalently with a quantum dot is considered. Before the laser pulse is switched on the time step can be larger due to the simple unitary time evolution. During the pulse the time step must be reduced in order to be able to resolve at least the frequency of the pulse. After the pulse the system is in an excited state, but the effective Hamiltonian is time-independent and the time step might again increase. In particular for this scenario we developed a method to adapt the time step for the described predictor-corrector scheme. We measure the absolute value of the maximal component of $\mathbf{Y}_i^{(j)} - \mathbf{Y}_i^{(j-1)}$ and stop the iterations if it is less than an appropriate bound, typically 10^{-10}. If the number of iterations is high the results of the predictor are not good enough and too much computation time is spent for the correction process. If the number of iterations is low, because the predicted result is particularly good, the corrector might not unfold its entire power. Even though the stop criterion for the iterations is met, the final result will not fit the precision required. Based on these considerations we establish the first criterion for the adaption of the time step: If the number of iterations is less than a lower bound (2) or greater than an upper bound (5) the time step might be adapted (numbers in brackets are typical values). The variation of the norm of the vector $\mathbf{y}_n$ is typically of the order of the machine precision. Thus, we measure errors in the norm in orders of the machine precision, which is for double precision floating point numbers $\epsilon_{mp} \approx 0.222 \times 10^{-15}$. We find the relative error in the norm $\sigma_n = (|\mathbf{y}_{n+1}| - |\mathbf{y}_n|)/\epsilon_{mp}$ and the absolute error in the norm $\eta_n = (1 - |\mathbf{y}_{n+1}|)/\epsilon_{mp}$ appropriate indicators for the adaption of the time step. The errors σ_n and η_n are integer numbers. If σ_n is too high (> 10) the complete propagation during the time step will be recalculated with a smaller time step. The bound for σ_n to decrease the time step without recalculation is lower (6). If σ_n is less than a lower bound (2), the time step will be increased, but only if the absolute error in the norm η_n is below a lower bound (10). Numbers in brackets give typical values for the error bounds. The described method for adaption of the time step has been tested successfully for a one-dimensional time-dependent problem of a particle in a Gaussian potential interacting with a laser pulse [166].

Finally, we use the predictor-corrector method with the same problem as before with the explicit methods (see Figure 4.3). We have tried the Fatunla method as a predictor, but the computation time increased beyond reasonable limits. For the Arnoldi method as predictor the computation time is comparable

Chapter 4. Time dependence and quantum fidelity

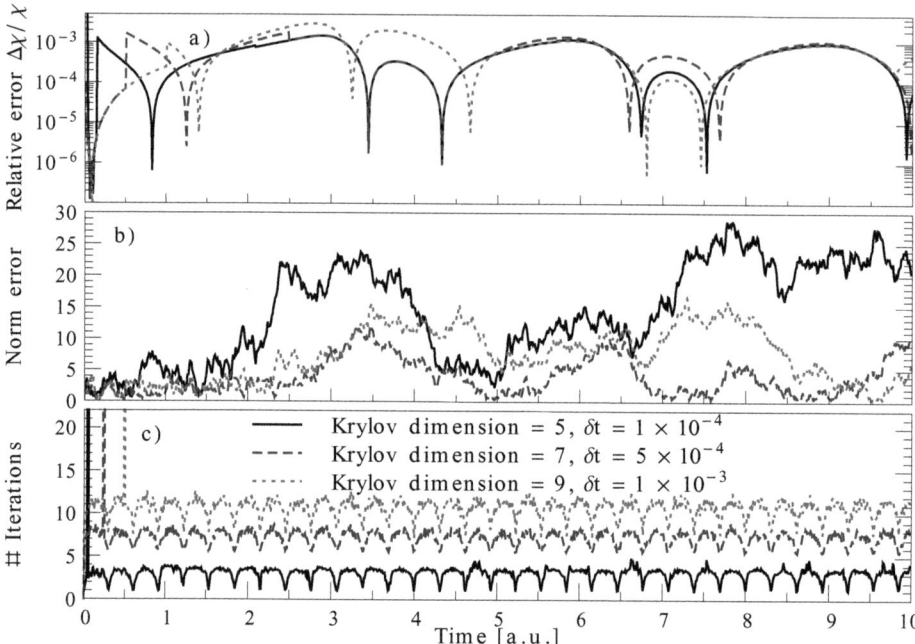

Figure 4.3: Properties of the predictor-corrector scheme calculating the quantum fidelity susceptibility of an eigenstate. The configuration coincides with Figures 4.1 and 4.2 except for the time evolution method. We show three results with different dimensions of the Krylov subspace and different time steps for the Arnoldi method as predictor (see legend in c)). The relative error of the susceptibility shown in a) is in average one order of magnitude smaller compared to the Arnoldi method. The error in the norm in b) is measured in units of the machine precision ϵ_{mp}. It is in average two orders of magnitude smaller than for the Arnoldi method. The number of iterations in c) determines the accuracy of the final result, in average more iterations lead to a better result. The time step remained constant, throughout all calculations, although it could have been adapted by the method. The calculation times are very long: 8 days, 5 hours, 43 minutes (dotted red line); 9 days, 17 hours, 14 minutes (dashed blue line); 17 days, 8 hours, 29 minutes (solid black line).

to the computation time for the pure Fatunla method. As a predictor we took the same three configurations for the Arnoldi method as before. The relative error for the susceptibility is at most below 10^{-3}, that is one order of magnitude better than the results with the Arnoldi method (see Figure 4.3 a)). However, it does not significantly depend on the configuration of the predictor. Surprisingly the accuracy in the norm depends on the configuration of the predictor, but the order is reversed (see Figure 4.3 b)). The best result was achieved with the lowest time step and the greatest Krylov dimension for the pure Arnoldi method. In the predictor corrector scheme the best result was achieved with the greatest time step and the lowest dimension of the Krylov subspace. Considering the number of iterations of the predictor-corrector scheme this corresponds to the argumentation for the first criterion for the adaption of the time step (see Figure 4.3 c)). As long as the predictor is within a certain accuracy a worse prediction leads to more iterations and the corrector leads to a more accurate result. In contrast, a very accurate prediction leads only to few iterations and the corrector does not have the opportunity to provide an optimal result.

Although we used the method introduced to adapt the time step for the calculations presented in Figure 4.3 the time step did not change. The problem is not time-dependent and no adaption of the time step occurred for the parameters we had chosen. For test systems with time-dependent Hamiltonians the adaption of the time step has shown to be a powerful tool [166, 167]. In future applications we will take advantage of the adaptive time step for interactions of the quantum dot with a laser pulse.

4.3 Preliminary results for the quantum fidelity

As a first attempt to understand the unusual behaviour in the mean-field models from first principles we consider the quantum fidelity evolution of eigenstates in the Hamiltonian (1.7) for different configurations.

4.3.1 Quantum fidelity susceptibility for eigenstates

The quantum fidelity decay for eigenstates is calculated from the time evolution in the perturbed system. In particular for small perturbations, $\epsilon \leq 10^{-5}$, the decay of the quantum fidelity is dominated by the linear response and thus by the susceptibility. In the previous section we have shown that the analytical formula (4.8) is equivalent to the results obtained by explicitly calculating the time evolution. The deviations could be minimised for highly accurate numerical methods. Consequently, we calculate the quantum fidelity susceptibility for various eigenstates and different parameters of the system from the static properties by applying Equation (4.8). In particular we analysed two situations with a regular Hamiltonian, Hooke's atom and the purely quartic non-interacting oscillator. The unperturbed Hamiltonian $\boldsymbol{H}_0$ for the purely quartic oscillator reads

$$\boldsymbol{H}_0 = \sum_{j=1}^{2} \left[-\frac{1}{2} \nabla_j^2 + \kappa\, r_j^4 \right], \qquad (4.26)$$

where we set $\kappa = 0.1$. This gives an energy scale comparable to the scale in Hooke's atom with $\omega_0 = 1$. In each case we assumed the perturbation $\boldsymbol{V}$ to be the harmonic or the quartic confining potential or the Coulombic interparticle interaction,

$$\boldsymbol{V}_{\text{harm}} = \frac{1}{2}\left(r_1^2 + r_2^2\right), \qquad \boldsymbol{V}_{\text{quart}} = r_1^4 + r_2^4, \qquad \boldsymbol{V}_{\text{int}} = \frac{1}{r_{12}}, \qquad (4.27)$$

respectively. The frequencies involved in Equation (4.8) are the same for all three perturbations $(E_k - E_N)$, only the amplitudes differ depending on the different coupling between the eigenstates due to the perturbation $(\langle N|V|k\rangle)$.

For the two regular unperturbed Hamiltonians the harmonic perturbation does in neither case introduce complexity, since it does not break the separability. In Hooke's atom the quartic perturbation couples COM and relative motion and significantly changes the dynamics. In contrast, for the non-interacting quartic oscillator the Coulomb perturbation couples the two formerly independent particles. We have investigated the behaviour of the susceptibility for the lowest 200 states of each system for the time interval $t \in [0, 1000]$. For this purpose we used a total basis of 6370 basis states and approximately 1000 converged eigenvalues. The perturbations that do not significantly excite the dynamics show a very homogeneous behaviour for all eigenstates. The maximal amplitudes are of the same order of magnitude for all states considered and the dominating frequencies are similar. The perturbations breaking the separability show a significantly different behaviour. Susceptibilities of different states are dominated by different frequencies and the maximal amplitude is strongly varying. Perturbations breaking the separability strongly couple many states and therefore the number of effectively contributing frequencies rises dramatically.

In Figures 4.4 and 4.5 typical behaviours of the susceptibilities of some eigenstates for the two cases discussed are shown. The harmonic perturbation in c) (both Figures) shows the regular homogeneous behaviour for the susceptibilities in both cases. The same is true for the Coulomb perturbation in Hooke's

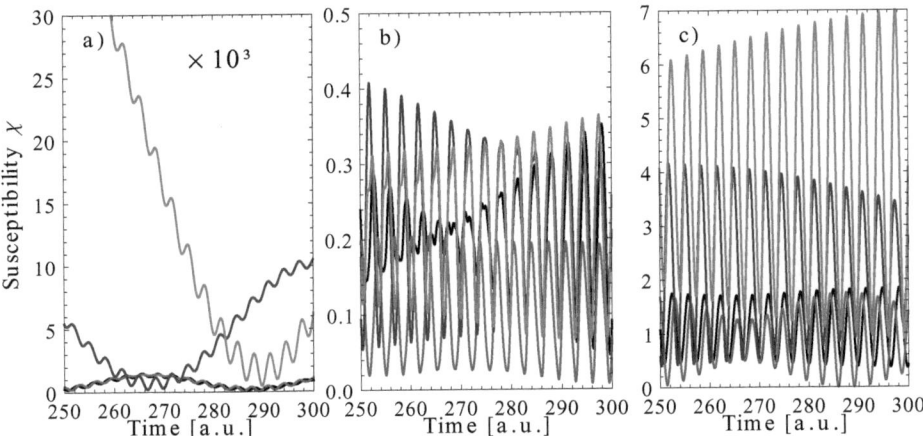

Figure 4.4: Quantum fidelity susceptibilities for four eigenstates in Hooke's atom with zero angular momentum, singlet spin and even parity. The states have the eigenenergies 4.846989 (black), 5.0 (red), 6.7530212, (blue) and 8.687964 (green). The susceptibilities are calculated in a basis of dimension 6370 by the formula (4.8). In a) the perturbation is quartic and therefore breaks the separability in COM and relative coordinates. The amplitudes of the susceptibilities are very high, due to the strong coupling between the eigenstates caused by the quartic potential. In b) the perturbation is the Coulomb interparticle interaction and keeps the regularity of the dynamics. The susceptibilities have a high degree of periodicity and the amplitudes vary only within a certain width. In c) the harmonic perturbation shows equivalent feature as the Coulomb perturbation.

atom, see Figure 4.4 b), and the quartic perturbation for the quartic oscillator, see Figure 4.5 a). The susceptibilities for Coulomb perturbation in the quartic oscillator, see Figure 4.5 b), shows the behaviour governed by various frequencies and higher amplitudes. Consider also the different scales for the Coulomb perturbation in both cases, while the eigenenergies of the states are almost equal. Equivalently, the quartic perturbation introduces very high amplitudes for Hooke's atom, see 4.4 a).

The behaviour of the quantum fidelity susceptibility for eigenstates shows characteristic features. These features can be used to distinguish between perturbations introducing complexity to the system and those that preserve the regularity. We have also calculated the susceptibilities of eigenstates for the same perturbations as above in the full potential ($\omega_0 = 1$, $\kappa = 0$, $\gamma = 1$) when the dynamics is chaotic before the perturbation. There are no significant differences in the behaviours for different perturbations. Furthermore, the susceptibilities neither show the high regularity of the regular perturbations, nor

Chapter 4. Time dependence and quantum fidelity

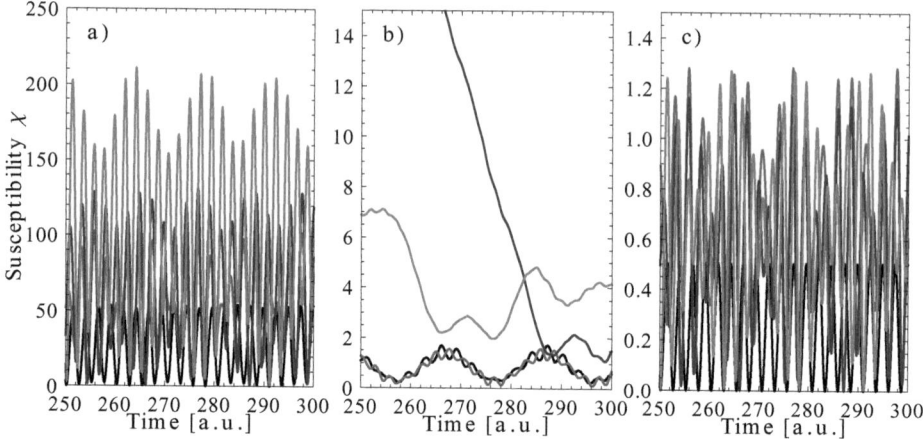

Figure 4.5: Quantum fidelity susceptibilities of four eigenstates in the purely quartic oscillator with $\kappa = 0.1$, zero angular momentum, singlet spin and even parity. The states have the eigenenergies 3.154563 (black), 3.472157 (red), 5.573050, (blue) and 8.076786 (green) and are comparable to the eigenenergies of Hooke's atom in Figure 4.4. The quartic perturbation leads to the regular behaviour of the susceptibilities in a). The amplitudes compared to Hooke's atom are two orders of magnitude smaller. In this case the Coulomb perturbation breaks the separability in two independent particles and the susceptibilities in b) show the more complex behaviour associated to stronger coupling between the eigenstates caused by the perturbing potential. The expected highly periodic homogeneous behaviour of the susceptibilities is shown in c) for a harmonic perturbation.

the high amplitudes and large number of relevant frequencies as in the cases breaking the separability. Since all perturbing potentials already contribute to the full potential no new couplings between the states are introduced by the perturbation, explaining the behaviour of the susceptibilities.

4.4 Summary

In this chapter we have introduced the concept of quantum fidelity as a dynamic measure of complexity. For the linear approximation of the fidelity, described by the susceptibility, we derived an analytical formula for initial eigenstates. This depends only on properties of the static system and we used it to test three different numerical methods for the time evolution. The method based on ideas of Fatunla was shown to be a sophisticated method with adaption of the time step, but with a lack in accuracy, in particular for our purpose. A second method, working in a Krylov subspace based on the matrix diagonalisation of Arnoldi, is a unitary time evolution method. Short computation times and the conservation of the norm make this method particularly suited for our purpose. For future applications with time-dependent Hamiltonians we described an iterative predictor-corrector scheme for the time evolution. We have presented an adaption of the time step for this scheme based on the accuracy of the norm.

In the second part we presented some results for the quantum fidelity moving towards the clarification of an unusual behaviour found in many-body systems. These systems have similar properties as our model system, but are treated in the mean-field approximation in contrast to our exact approach. As a first step we investigated the quantum fidelity susceptibility for eigenstates and found it capable of distinguishing between perturbations that leave the dynamics regular and those that break the separability. To some extent this is related to the original idea of Peres to have a dynamical measure of quantum chaos similar to the property of classical chaos. Indeed, the susceptibility was calculated purely by the properties of the static system and the dynamic aspect can only be regarded as a tool to reveal the complexity hidden in the spectrum of the static system.

The unusual behaviour of the quantum fidelity decay found in mean-field models of many-body systems has not been confirmed for eigenstates. All eigenstates show nearly complete revivals for weak perturbations, corresponding to the susceptibility assuming the value zero. Preliminary results for centred Gaussian initial states show the usual decay behaviour. The investigations for different regimes of the potential and differently shaped Gaussian initial states is still in progress. Thus, it can not be excluded that the unusual behaviour might be confirmed with our model. As an alternative future perspective to clarify the unusual behaviour we wish to relate it to quantum freezing.

Chapter 5
Hund's first multiplicity rule

Hund's first rule (HFR) [16, 17] is well-known to have universal predictive power for the ground-state electron configuration of atoms: "Among the many-electron states arising from the same configuration, the ground state has the largest total spin S." The rule can be generalised to excited states with the same configuration of one-electron orbitals: "Among the many-electron states arising from the same configuration of one-electron orbitals, the energy of the state decreases as the total spin S of the state increases" [168, 25].

There has been a long discussion about the origin of this effect, starting from the first explanation by Slater [169]. He claimed that the difference in Coulomb interaction between the electrons due to the existence of a Fermi hole was responsible for this effect. The Fermi hole describes the minimum of the probability density of the spatial wave function at the origin, due to the antisymmetrisation for fermionic states. It was Slaters conviction, that his concept of antisymmetrisation (Slater's determinant[1]) could fully account for the origin of HFR. More than thirty years later a numerical calculation by Davidson [172] for atomic helium was the first to reveal that Slater's argumentation was wrong; in fact the Coulomb interaction was higher for the triplet state. Several calculations for atomic helium and helium-like ions supported this result [173, 174, 175, 176]. Calculations for atomic systems with more electrons, e.g. Ref. [177], confirmed this behaviour. A result based on perturbation theory [178, 179] showed that generally all atoms violate Slater's explanation and a new interpretation was given by Boyd [180]. His argumentation follows the virial theorem for atomic systems: the total energy equals half of the potential energy. All interactions are Coulombic and the potential energy is the sum of the inter-electron and the negative electron-nucleus interactions. As a consequence the absolute value of the nuclear potential energy must rise for higher

---
[1]Though it must be stated that the formal determinantal treatment was introduced independently by Heisenberg [170] and by Dirac [171] and not by Slater.

spin states to compensate the higher inter-electron interaction. This could be interpreted as a higher effective nuclear charge or lower screening of the nuclear attraction, caused by different angular configurations. Only recently this interpretation has again been doubted by Sajeev et al. [22] when they compared the effect in atoms to rectangular quantum dots. They claim that the angular configuration does not play a major role for screening effects. A generalisation of HFR for configurations with two electrons in two open shells ($m_1 \geq 1$, and $m_2 \geq 1$), distinguishing different behaviours between natural and unnatural parity states, was first proposed by Russell and Meggers [181]. A precise formulation in terms of the independent particle model by Morgan and Kutzelnigg [168] is referred to as the *alternating rule*. A very sound review on Hund's rules in atomic systems has also been presented by these authors [25].

5.1 Hund's first rule in quantum dots

Just like in atoms, the filling of the electron orbitals for few-electron quantum dots in the ground state is in general governed by Hund's rules [18, 20, 21]. For different quantum dot models it was found, that Slater's explanation holds, in contrast to atomic systems [22, 106, 24].

One of the first results on quantum dots considers two-electron spherically-symmetric rectangular quantum dots [22]. There it was found that HFR is followed for the $^{1/3}S(1s2s)$ state[2] and can be explained by the difference in Coulomb repulsion between the electrons. The authors also find that the origin of HFR does not depend on the angular configuration, for rectangular quantum dots, as well as for atomic helium. That is a remarkable result, as it contradicts the established interpretation for atomic systems by Boyd [180].

The application of HFR to Hooke's atom has been addressed by Sako et al. [106] and has also been related to the case of atomic helium [23]. By transforming to COM and relative coordinates they find corresponding angular quantum numbers (m_c and m_r) and the restriction for m_r to be even (odd) for singlet (triplet) states. These angular momentum quantum numbers are responsible for the localisation properties in the radial coordinates R_{COM} and r_{rel} and, thus, for the difference in Coulomb repulsion. Unfortunately, their exposure is rather unclear and the localisation properties are not significantly visualised. Furthermore they claim, that several pairs of singlet and triplet

---

[2]In our notation these states correspond to $|0,0,1,0\rangle_{idp}^{\epsilon_s=\pm 1, \epsilon_p=\pm 1}$, although we believe, that only even parity states were considered, since a different behaviour can be expected from our results on odd parity states for a similar case, see Section 5.4.2.

states can be compared according to HFR, which is wrong, as we will show in Section 5.2.1. Still the central feature, the interplay between the angular quantum numbers m_c and m_r and the localisation of the state, will guide us to the non-separable cases of the quartic oscillator and further.

Another recent approach addresses the atomic helium isoelectronic sequence and a model of Hooke's atom in three dimensions [24]. As a consequence of the virial theorem and a perturbative treatment of the Coulomb repulsion between the electrons, V_{Coulomb}, an asymptotic value, in the non-interacting limit, is found for the ratio

$$\lambda_{\text{confinement}} = \frac{\langle \Delta V_{\text{Coulomb}} \rangle}{\langle \Delta V_{\text{confinement}} \rangle}, \tag{5.1}$$

where $V_{\text{confinement}}$ is the confining potential. The Δ denotes the difference taken between the potential energies of corresponding singlet and triplet states. As already indicated by the index, $\lambda_{\text{confinement}}$ depends explicitly on the confining potential. For the isoelectronic sequence of atomic helium the non-interacting limit is reached for high nuclear charge. Besides the value for harmonic confinement stated in [24], we provide the general result for a positive-power-law potential $V_n \propto r^n$ to be

$$\lambda_n = \frac{\langle \Delta V_{\text{Coulomb}} \rangle}{\langle \Delta V_n \rangle} = n + 2. \tag{5.2}$$

The greater the exponent of the power-law potential, the more is the difference between singlet and triplet states dominated by the difference in Coulomb interaction. However, also in [24] no analysis on the applicability of HFR for the harmonic confinement in three dimensions is performed. From our results for the planar case, we believe that the applicability is, here as well, restricted to a limited number of states.

5.2 Harmonic case

Since the existing investigations on HFR in Hooke's atom are, in our opinion, incomplete, we will address this issue carefully. We will focus on the question if HFR is applicable at all, an ambiguous concern due to the separability in COM and relative coordinates. As a preliminary step in order to discuss the applicability we will focus on the topic of entanglement and put our results in correlation with recent literature. Indeed we find that HFR is not applicable for the planar model, except for one particular case, which includes both parities. We will then focus on this individual case and use it as an illustrative example.

Chapter 5. Hund's first multiplicity rule

Along the way we will introduce basic concepts that will assist us to clarify the origin of the *alternating rule* in more general confinements.

5.2.1 Entanglement and separability

Entanglement is a term used to describe quantum correlations between the particles. In the well-known case of systems with distinguishable particles the Schmidt rank determines the entanglement of the system. A general bipartite pure state $|\Phi\rangle$ can be expressed as the sum over tensor products of the respective basis-states,

$$|\Phi\rangle = \sum_{j,k} c_{j,k} |\varphi_j\rangle \otimes |\phi_k\rangle. \qquad (5.3)$$

This expression is called the Schmidt decomposition in the orthonormal bases $\{|\varphi_j\rangle\}$ and $\{|\phi_k\rangle\}$. The Schmidt rank is the minimal number of non-vanishing coefficients $c_{j,k}$, consequently a non-entangled state has Schmidt rank unity. A convenient tool to analyse entanglement properties is the reduced density matrix

$$\rho_{\text{red}} = \sum_j \langle \varphi_j | \Phi \rangle \langle \Phi | \varphi_j \rangle, \qquad (5.4)$$

which is obtained by performing the partial trace over one of the subsystems. The particular choice of subsystem over which the density matrix is reduced is arbitrary and does not affect the final result. The Schmidt rank equals the rank of the reduced density matrix. In lieu of determining the rank directly entanglement witnesses have been introduced, of which the von-Neumann entropy $S[\rho_{\text{red}}]$ of the reduced density matrix is one of the most commonly used

$$S[\rho_{\text{red}}] = -\text{Tr}[\rho_{\text{red}} \log_2(\rho_{\text{red}})]. \qquad (5.5)$$

The von-Neumann entropy of a non-entangled state equals zero, while it is greater than zero for an entangled state. For further details see topical reviews on this matter, e.g. [38, 39, 182].

The situation is more intricate for indistinguishable particles, where a linear superposition is generally incorporated for all states. A clear description of the bosonic and fermionic entanglement properties has been given in [183, 184]. For fermionic states the Slater rank is introduced, which is the minimal number of non-vanishing coefficients in the Slater decomposition

$$|\Psi\rangle = \sum_{j,k} c_{j,k} \left(|\psi_{1,j}\rangle \otimes |\psi_{2,k}\rangle - |\psi_{2,j}\rangle \otimes |\psi_{1,k}\rangle \right). \qquad (5.6)$$

5.2. Harmonic case

Here, $\{|\psi_{n,j}\rangle\}$ is a common orthonormal basis for both particles, where n denotes the particle. If the Slater rank of a quantum state is unity it is a non-entangled state, i.e. the only correlations that exist between the fermions can be attributed to their indistinguishable nature. In order to determine the Slater rank the partial trace over one of the particles is performed on the density matrix, which defines the reduced density matrix,

$$\rho_{\text{red}} = \text{Tr}_2 [|\Psi\rangle\langle\Psi|] = \sum_j \langle\psi_{2,j}|\Psi\rangle\langle\Psi|\psi_{2,j}\rangle. \tag{5.7}$$

We consider two entanglement witnesses, the reduced von-Neumann entropy $\mathcal{E}_{\text{VN}}$ and the reduced linear entropy $\mathcal{E}_{\text{L}}$ for pure states of two identical fermions,

$$\mathcal{E}_{\text{VN}} = S[\rho_{\text{red}}] - \log_2 2, \tag{5.8}$$
$$\mathcal{E}_{\text{L}} = 1 - 2\,\text{Tr}[\rho_{\text{red}}^2]. \tag{5.9}$$

Cancelling the amount of entropy corresponding to the antisymmetrisation of the fermionic states ($\log_2 2$ and 1, respectively) both measures vanish if and only if a state is non-entangled in the fermionic sense [185]. The advantage of (5.9) is that the reduced density matrix needs not to be diagonalised for evaluation. Therefore, it is most commonly used in numerical treatments.

Up to now we have considered a general state $|\Psi\rangle$, describing two fermions, that can be written as a Slater determinant. Turning on an interaction between the fermions introduces entanglement. We will only consider interactions that do not explicitly couple the spatial and the spin degrees of freedom. Thus, entanglement stems from the separability inherited by this kind of interaction, and remains, even in the non-interacting limit, as an *offset entanglement*. As explained in Section 1.1.2 the spins couple to give singlet and triplet states due to the interaction. The separability of the Hamiltonian is carried forward to a product wave function and further to a product of reduced density matrices,

$$\rho_{\text{red}} = \rho_{\text{red}}^{\text{spatial}}\, \rho_{\text{red}}^{\text{spin}}. \tag{5.10}$$

Notice, that this holds only since the one-particle basis, used to trace over the density matrix, can equivalently be expressed as product states $|\psi_{2,j}\rangle = |\psi_j(\mathbf{r}_2)\rangle \otimes |(S, S_z)_{2,j}\rangle$ of a spatial- and a spin-dependent state. Plugging (5.10) into (5.8) and (5.9) we find

$$\mathcal{E}_{\text{VN}} = S[\rho_{\text{red}}^{\text{spatial}}] + S[\rho_{\text{red}}^{\text{spin}}] - 1, \tag{5.11}$$
$$\mathcal{E}_{\text{L}} = 1 - 2\,\text{Tr}[(\rho_{\text{red}}^{\text{spatial}})^2]\,\text{Tr}[(\rho_{\text{red}}^{\text{spin}})^2]. \tag{5.12}$$

Chapter 5. Hund's first multiplicity rule

The spin-dependent parts are easily evaluated for the spin states $|S, S_z\rangle$,

$$S[\rho_{\text{red}}^{\text{spin}}(|0,\ 0\rangle)] = \qquad S[\rho_{\text{red}}^{\text{spin}}(|1,\ 0\rangle)] = 1, \qquad (5.13)$$

$$S[\rho_{\text{red}}^{\text{spin}}(|1,+1\rangle)] = \qquad S[\rho_{\text{red}}^{\text{spin}}(|1,-1\rangle)] = 0, \qquad (5.14)$$

$$\text{Tr}[(\rho_{\text{red}}^{\text{spin}}(|0,\ 0\rangle))^2] = \qquad \text{Tr}[(\rho_{\text{red}}^{\text{spin}}(|1,\ 0\rangle))^2] = \frac{1}{2}, \qquad (5.15)$$

$$\text{Tr}[(\rho_{\text{red}}^{\text{spin}}(|1,+1\rangle))^2] = \qquad \text{Tr}[(\rho_{\text{red}}^{\text{spin}}(|1,-1\rangle))^2] = 1, \qquad (5.16)$$

and depends only on $|S_z|$. By considering the interparticle interaction with first order perturbation theory for a fourfold degenerate subsystem the authors of [47] find this kind of entanglement. Independently of the confining potential these states will be entangled for $S_z = 0$. However, this effect does not explain the non-vanishing values of entanglement in atomic models with harmonic confinement for decreasing interaction strength, as they claim. Indeed, another separability induces this non-vanishing entanglement.

For two distance-dependently interacting fermions in harmonic confinement the system is separable in COM and relative motion. This holds in any dimension, also for anisotropic harmonic potentials. The separability leads to the eigenstates being product states of COM and relative wave functions, which can preliminarily be incorporated. Regarding the interaction as a perturbation the correct symmetrisation of the eigenstates can be found by diagonalising the interaction matrix in a basis of degenerate eigenstates [47]. Entanglement is introduced by the symmetrisation already before the interaction actually couples the two particles. As a consequence the entanglement for such systems will not generally vanish in the non-interacting limit. This explains the behaviours that were reported for several harmonically confined models. The explicit kind of inter-electron interaction does not play a major role, as long as it depends solely on the interparticle distance. With this we include the Crandall atom, with inverse square interaction, and Hooke's atom discussed in [46], as well as the Moshinsky atom, with harmonic interaction, and a model with contact interaction addressed in [47].

Systems without explicit separability in COM and relative coordinates usually relax in a basis of independent particles and will not show such an offset entanglement. The linear entropy of atomic helium has been shown to saturate to the value of one half [47, 45]. In contrast, the linear entropy for states in quantum dot models with harmonic confinement saturates to the higher value of unity [47]. For most of the states the entropies in the non-interacting case already exceed the limit for atomic helium, see Figure 5.1. The offset entanglement is thus responsible for the higher limit for the linear entropy in the harmonic cases. Furthermore the linear entropy for the interacting model has been shown to increase with the energy [47]. This dependence is already contained in the non-interacting limit (see Figure 5.1).

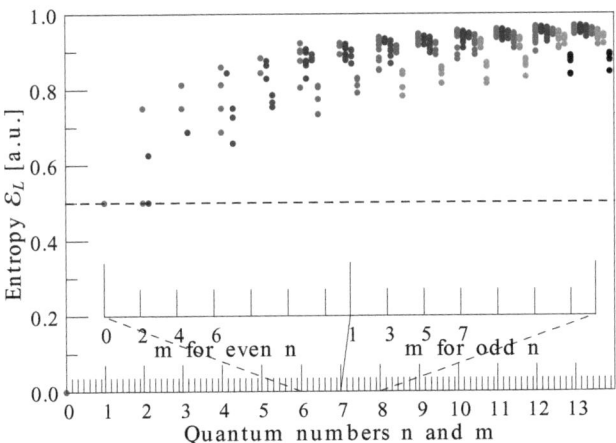

Figure 5.1: Linear entropies $\mathcal{E}_\mathrm{L}$ for all eigenstates in the non-interacting harmonic model up to principal quantum number $n = 13$. Points corresponding to the same principal quantum number n, but different angular momenta m are displaced horizontally for better visibility, as indicated for $n = 6$ and 7 by the enlarged scale. The spin quantum number is set to $S_z \equiv 0$ for all states. For non-vanishing angular momentum we consider states without parity symmetrisation. The dashed horizontal line corresponds to the saturation limit for the linear entropy in atomic helium. The eigenenergies are given by $E_n = \omega_0(n+2)$ a.u..

Another effect can be interpreted in terms of the offset entanglement that emerges from some separability of the system considered. For an anisotropic harmonic system with $\omega_z \gg \omega_0 = \omega_x = \omega_y$ and Coulomb interparticle interaction a magnetic field is applied [42]. As shown in [101] the primarily cylindrically symmetric system becomes spherically symmetric for magnetic fields, such that $\omega_z \equiv \sqrt{\omega_B^2 + \omega_0^2}$, with the Larmor frequency ω_B. At this specific value a minimum of the entanglement is found for several different strengths of the confinement parameter ω_0. The states in the non-interacting limit belong either to a symmetry class of the cylindrical or of the spherical symmetry. Qualitatively speaking the interaction affects the states differently according to their algebraic properties. We expect that the offset entanglement is different, depending on the symmetry. In particular, we expect the spherically symmetric representation to be less entangled than the cylindrically symmetric one. This discontinuity in the non-interacting limit carries forward to a minimum in the

Chapter 5. Hund's first multiplicity rule

interacting case.

Table 5.1: Entanglement witnesses for low-lying states in the non-interacting harmonic model. States are identified with the basis representation in COM and relative coordinates with parity symmetrisation. We show two entanglement witnesses, the reduced von-Neumann entropy $\mathcal{E}_{\text{VN}}$ and the reduced linear entropy $\mathcal{E}_{\text{L}}$, which are zero for non-entangled states and greater than zero for entangled states. A fermionic state is not entangled, if it is the result of antisymmetrising two orthogonal one-particle states. The entropies include the entanglement introduced by the spin dependent part of the wave function, which depends only on $|S_z|$. Energies are given in modified atomic units for $\omega_0 = 1$, the other results are independent of the value of ω_0.

E	m	$\lvert n_c, m_c, n_r, m_r \rangle^{\epsilon_p}$	$\mathcal{E}_{\text{VN}}, S_z = 0$	$\mathcal{E}_{\text{VN}}, S_z = \pm 1$	$\mathcal{E}_{\text{L}}, S_z = 0$	$\mathcal{E}_{\text{L}}, S_z = \pm 1$
2	0	$\lvert 0,0,0,0 \rangle^{+1}$	0.0		0.0	
3	1	$\lvert 0,1,0,0 \rangle^{\pm 1}$	1.0		0.5	
3	1	$\lvert 0,0,0,1 \rangle^{\pm 1}$	1.0	0.0	0.5	0.0
4	0	$\lvert 1,0,0,0 \rangle^{+1}$	2.0		0.75	
4	0	$\lvert 0,0,1,0 \rangle^{+1}$	2.0		0.75	
4	0	$\lvert 0,1,0,-1 \rangle^{\pm 1}$	1.0	0.0	0.5	0.0
4	2	$\lvert 0,2,0,0 \rangle^{\pm 1}$	2.0		0.75	
4	2	$\lvert 0,0,0,2 \rangle^{\pm 1}$	2.0		0.75	
4	2	$\lvert 0,1,0,1 \rangle^{\pm 1}$	1.0	0.0	0.5	0.0
5	1	$\lvert 0,1,0,-2 \rangle^{\pm 1}$	2.29879	1.29879	0.65625	0.53125
5	1	$\lvert 0,1,1,0 \rangle^{\pm 1}$	2.5		0.8125	
5	1	$\lvert 1,1,0,0 \rangle^{\pm 1}$	2.29879		0.65625	
5	1	$\lvert 0,0,1,1 \rangle^{\pm 1}$	2.29879	1.29879	0.65625	0.53125
5	1	$\lvert 1,0,0,1 \rangle^{\pm 1}$	2.5	1.5	0.8125	0.625
5	1	$\lvert 0,2,0,-1 \rangle^{\pm 1}$	2.29879	1.29879	0.65625	0.53125
5	3	$\lvert 0,1,0,2 \rangle^{\pm 1}$	2.06128		0.703125	
5	3	$\lvert 0,3,0,0 \rangle^{\pm 1}$	2.56128		0.828125	
5	3	$\lvert 0,0,0,3 \rangle^{\pm 1}$	2.56128	1.56128	0.828125	0.65625
5	3	$\lvert 0,2,0,1 \rangle^{\pm 1}$	2.06128	1.06128	0.703125	0.40625

In order to clarify the question on the applicability of HFR in the next section and to corroborate our assumptions about general harmonic confinement we have explicitly calculated the entanglement for our non-interacting harmonic model. The results for the entanglement witnesses $\mathcal{E}_{\text{VN}}$ and $\mathcal{E}_{\text{L}}$ for both possible values of the spin z-component are given in Table 5.1. There are only few non-entangled states, the only non-entangled singlet state is the ground state, while the only non-entangled triplet states are those with $E = 3$ and $E = 4$ and

$S_z = \pm 1$. We have analysed all states up to $E = 15$ and all symmetry classes and angular momenta. For the sake of completeness we also applied our formalism to states with positive and negative non-vanishing angular momentum. This change of symmetrisation does not alter our results qualitatively, compare Figure 5.1. For $\gamma = 0$ the entanglement between the two fermionic particles in terms of the quantum numbers describing the state $|n_c, m_c, n_r, m_r\rangle^{\epsilon_p}$ can be calculated analytically.

We consider the density matrix

$$|\Psi\rangle\langle\Psi| = |n_c, m_c, n_r, m_r\rangle^{\epsilon_p} {}^{\epsilon_p}\langle n_c, m_c, n_r, m_r|, \qquad (5.17)$$

and represent it in an independent particle basis. Thus, we transform to a basis associated with Cartesian coordinates $|C_x, C_y, R_x, R_y\rangle^{\epsilon_p} = |n_c, m_c, n_r, m_r\rangle^{\epsilon_p}$. The Cartesian basis in COM (C_x, C_y) and relative (R_x, R_y) coordinates is defined by:

$$|C_x, C_y, R_x, R_y\rangle^{\epsilon_p} =$$
$$= \begin{cases} |C_x, C_y, R_x, R_y\rangle & \text{if } C_x = C_y \wedge R_x = R_y, \\ (|C_x, C_y, R_x, R_y\rangle + \epsilon_p |C_y, C_x, R_y, R_x\rangle)/\sqrt{2} & \text{otherwise.} \end{cases}$$

The transformation is expressed through

$$n_c = \min(C_x, C_y), \qquad m_c = |C_x - C_y|, \qquad (5.18)$$
$$n_r = \min(R_x, R_y), \qquad m_r = \pm|R_x - R_y|, \qquad (5.19)$$

such that $m_c + m_r = C_x - C_y + R_x - R_y$. It is easy to evaluate the transformation into the Cartesian independent particle basis $|n_x, n_y, k_x, k_y\rangle$ making use of the simple harmonic oscillator operator algebra and a symmetric coordinate transformation:

$$x_C = \frac{1}{\sqrt{2}}(x_1 + x_2), \quad y_C = \frac{1}{\sqrt{2}}(y_1 + y_2), \quad x_R = \frac{1}{\sqrt{2}}(x_1 - x_2), \quad y_R = \frac{1}{\sqrt{2}}(y_1 - y_2).$$

The transformation matrix elements are given by

$$\langle n_x, n_y, k_x, k_y | C_x, C_y, R_x, R_y \rangle = \sqrt{\frac{n_x! n_y! k_x! k_y!}{C_x! C_y! R_x! R_y!}} \sqrt{1/2}^{(n_x + n_y + k_x + k_y)}$$
$$\times F(C_x, R_x, k_x)\, \delta_{C_x + R_x, n_x + k_x} F(C_y, R_y, k_y)\, \delta_{C_y + R_y, n_y + k_y}, \qquad (5.20)$$

with the function

$$F(N, M, k) := \sum_{r=\max(0, k-N)}^{\min(k, M)} \binom{N}{k-r}\binom{M}{k}(-1)^r. \qquad (5.21)$$

Chapter 5. Hund's first multiplicity rule

Applying Equation (5.7) we obtain the reduced density matrix. Evaluation of the entanglement witnesses (5.8) and (5.9) boils down to the expressions

$$\mathcal{E}_{\text{VN}} = -\sum_j (\lambda_j \log_2 \lambda_j) + S[\rho_{\text{red}}^{\text{spin}}] - 1, \qquad (5.22)$$

$$\mathcal{E}_{\text{L}} = 1 - \text{Tr}[(\rho_{\text{red}}^{\text{spin}})^2] \sum_j (\lambda_j^2), \qquad (5.23)$$

with λ_j the eigenvalues of ρ_{red}.

5.2.2 Applicability of Hund's rule

Understanding the applicability of HFR for the harmonic case requires the identification of states that arise from the same configuration of one-electron orbitals. This, however, is not a trivial task due to the separability in COM and relative coordinates. The spatial fermionic entanglement for indistinguishable particles is particularly suited for this purpose. At this point it is necessary to clearly distinguish our application from the presented formalism and the general term of entanglement. Our goal is to identify states that are the direct (anti-)symmetrisation of one-particle orbitals, as stated in the empiric definition of HFR [168]. This rule is valid in the non-relativistic limit, where no coupling between spatial and spin degrees of freedom is presumed. Consequently, we do not consider the spin-dependent part of the wave function at all, as it has no influence on the energy of the state. The difference is easily explained with an example. The state $|0, 1, 0, 0\rangle^{+1}$ can be expressed with a Slater decomposition of rank unity and gives $\mathcal{E}_{\text{VN}} = 0$ in the non-interacting limit. Thus, it is not spatially entangled. Taking into consideration also the spin-dependent wave function, which is singlet, the state is indeed entangled in the fermionic sense, since

$$\mathcal{E}_{\text{VN}}\left[|0, 1, 0, 0\rangle^{+1} \otimes |0, 0\rangle\right] = 1. \qquad (5.24)$$

The corresponding state of atomic helium is the $(1s2p)\,^1P$ state, for which HFR is applicable in combination with the triplet state $(1s2p)\,^3P$. A unique case is the ground state, for which we find $\mathcal{E}_{\text{VN}}\left[|0, 0, 0, 0\rangle^{+1}\right] = -1$, since the symmetrisation is exclusively contained in the spin wave function. This will be of no further concern to us: there is no corresponding triplet state for the ground state.

In the previous section we explained, that already the choice of the appropriate symmetry class creates entanglement of the states. The symmetry and the exact quantum numbers n_c, m_c, n_r and m_r are independent of the value of the interaction strength γ. We have thus found a possible choice of symmetry class induced by the interaction. For non-vanishing angular momenta, omitting the parity and allowing negative values for m offers another choice of an appropriate basis.

In Table 5.1 and Figure 5.1 we show the reduced von-Neumann entropy $\mathcal{E}_{\text{VN}}$ and the reduced linear entropy $\mathcal{E}_{\text{L}}$ for various eigenstates. The information about the spatial entanglement of a state can be acquired by subtracting unity from $\mathcal{E}_{\text{VN}}$ for $S_z = 0$. The only pairs of spatially non-entangled singlet and triplet states we found are $|0,1,0,0\rangle^{\epsilon_p}$ and $|0,0,0,1\rangle^{\epsilon_p}$. Here the even- and odd-parity states are degenerate, so we only consider the even case $\epsilon_p = 1$. For the second choice of the basis the only spatially non-entangled states we found are the corresponding pairs $(|0,1,0,0\rangle, |0,0,0,1\rangle)$ with $m = 1$ and $(|0,-1,0,0\rangle, |0,0,0,-1\rangle)$ with $m = -1$.

5.2.3 Origin of Hund's rule

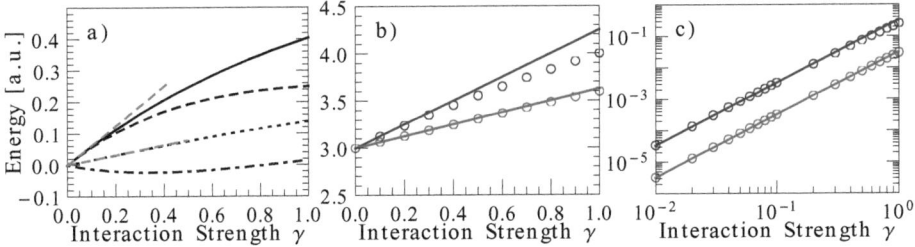

Figure 5.2: Adiabatically raising the interparticle interaction via the Coulomb strength parameter γ lifts the degeneracy of the singlet state $|0,1,0,0\rangle^{+1}$ and the triplet state $|0,0,0,1\rangle^{+1}$ in the harmonic confining potential. The singlet state (blue) has higher energy than the triplet state (red). In a) the differences between the singlet and the triplet energy contributions, according to the virial theorem (2.36), total energy (solid line), Coulomb interaction (dashed line), harmonic potential (dotted line) and kinetic energy (dot-dashed line), are shown. The dashed green lines are the linear fits in the non-interacting limit for the Coulomb interaction and the harmonic confinement. The ratio $\lambda_2 = 3.9999995$ of the two slopes obtained for $\gamma = 10^{-7}$ confirms the analytical result (5.2). In b) numerically exact eigenenergies E_{num} (circles) and the results from first-order perturbation theory E^1 (lines) are shown. The deviations $E^1 - E_{\text{num}}$ (circles) are shown in c) and are close to the quadratic fit for the triplet and the singlet case. Small deviations are observed close to $\gamma = 1$. The coefficient of the quadratic fit of the singlet state is larger by a factor of approximately ten, in accordance with the significant change in the wave function (compare Figure 5.3).

For the unique pair of singlet and triplet states in Hooke's atom to which HFR can be applied the question about validity and its origin emerges. These

Chapter 5. Hund's first multiplicity rule

states follow HFR and the singlet energy grows larger than the corresponding triplet energy with adiabatically increasing interparticle interaction. We show the details of this effect by considering the different parts, which contribute to the total energy in Figure 5.2 a). We observe, that the total energy, the Coulomb energy and the confining harmonic energy are higher for the singlet state. The Coulomb interaction is the main contribution of the energy difference between singlet and triplet states. In contrast to the atomic case, Slater's explanation holds for the harmonic quantum dot: the Fermi hole causes the lower value of the triplet state dominated by the difference in Coulomb interaction. The result for the ratio $\lambda_2 = 3.9999995$ confirms the analytical value derived by Katriel [24].

In order to understand the origin of the Fermi hole we take a deeper look at the structure of the effective potential. This potential consists of the ordinary potential and the dynamic part coming from the angular motion which corresponds to the classical angular momentum barrier. The antisymmetry under interparticle exchange imposes even (odd) values of the angular momentum quantum number in relative coordinates, m_r, for the singlet (triplet) case. The effective potential depends on this quantum number and thus on the symmetrisation and, already for the non-interacting case, determines the localisation properties of the state. This localisation is responsible for the amount of influence of the Coulomb interaction. The singlet state has vanishing angular momentum in the relative coordinate, while for the triplet state $m_r = \pm 1$. In general it may well be possible, that a singlet state has higher relative angular momentum than the corresponding triplet state which would cause a violation of Hund's rule.

The localisation of these states, their corresponding effective potentials and influences of the Coulomb potential are illustrated in Figure 5.3. The singlet state localises close to the origin in the relative coordinate, because it does not feel an angular momentum barrier in this coordinate. The Coulomb term thus has a large influence on the singlet state and the expectation value of the Coulomb interaction reads

$$^{+1}\langle 0,1,0,0| \frac{1}{r_{\rm rel}} |0,1,0,0\rangle^{+1} = \sqrt{\frac{\pi}{2}} \text{ a.u.} . \qquad (5.25)$$

In addition the Coulomb barrier changes the effective potential and the localisation of the state. The triplet state has no probability of presence close to the origin of the relative coordinate, which is reflected in a smaller value of the

5.2. Harmonic case

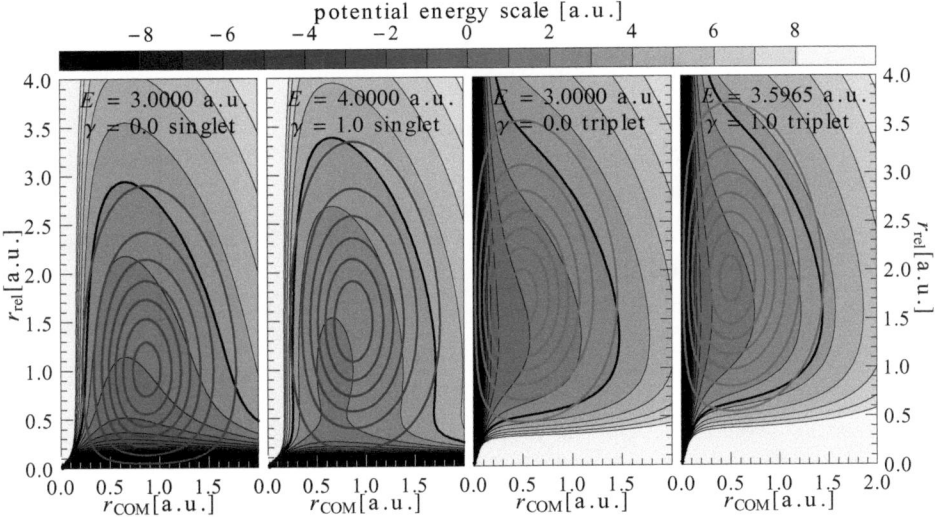

Figure 5.3: The probability densities of the singlet state $|0,1,0,0\rangle^{+1}$ (blue) and the triplet state $|0,0,0,1\rangle^{+1}$ (red) in the non-interacting ($\gamma = 0$, respectively left) and interacting ($\gamma = 1$, respectively right) harmonic case are pictured here. The grey shadings represent the confining effective potential with the thick equipotential line equal to the eigenenergy of the state. The coloured lines are contour lines of the probability density starting from 0.05 in steps of 0.1. The probability density of the singlet state is localised close to the origin of the relative coordinates for the non-interacting case and therefore significantly affected by the Coulomb interaction. The triplet state is localised close to the origin of the COM coordinate, but the angular momentum barrier in the relative coordinate repels the wave function from the origin. The Coulomb interaction causes only minor changes in the effective potential and in the localisation of the triplet state.

Coulomb interaction compared with the singlet case,

$$^{+1}\langle 0,0,0,1|\frac{1}{r_{\text{rel}}}|0,0,0,1\rangle^{+1} = \sqrt{\frac{\pi}{8}} \text{ a.u.}. \qquad (5.26)$$

In the interacting case only minor changes in the effective potential and the localisation can be observed.

The rather low dependence of the localisation of both states on the Coulomb interaction suggests application of first-order perturbation theory. The energy levels depend nearly linearly on γ, see Figure 5.2 b), particularly for the triplet state, and the deviation is quadratic in γ (see Figure 5.2 c)). They are well described with second-order perturbation theory for both cases.

Chapter 5. Hund's first multiplicity rule

5.3 Quartic case

For purely quartic confining potential the Hamiltonian (1.7) reduces to

$$H_{\text{quartic}} = \sum_{j=1}^{2}\left[-\frac{1}{2}\nabla_j^2 + \kappa\, r_j^4\right] + \frac{\gamma}{r_{12}}, \qquad (5.27)$$

where we set $\kappa \equiv 1$, the separability in COM and relative motion is lost. Still we can express the potential in the coordinates r_{COM}, r_{rel} and $\varphi = \varphi_{\text{COM}} - \varphi_{\text{rel}}$, such that the confining potential is given by

$$V_{\text{quartic}}(r_{\text{COM}}, r_{\text{rel}}, \varphi) = 2\,r_{\text{COM}}^4 + \frac{1}{8}r_{\text{rel}}^4 + r_{\text{COM}}^2 r_{\text{rel}}^2 (3 - 2\sin^2\varphi). \qquad (5.28)$$

There is no dependence on the angle $\vartheta = \varphi_{\text{COM}} + \varphi_{\text{rel}}$, which is the conjugate coordinate to the conserved total angular momentum perpendicular to the plane.

For potentials lacking a further separation, apart from the separation in independent particles for the non-interacting case, there is no subtlety about the applicability of HFR. The interaction does not induce a symmetry other then the independent particle basis and no offset entanglement is introduced to the spatial wave function. This is also valid for the quartic potential. All states arise from (anti-)symmetrisation of one-particle orbitals and the non-interacting eigenenergies can be expressed as the sum of the one-particle solutions, see Section 2.2.3.

Table 5.2: Numerically obtained properties of the states $|1,1,0,-1\rangle_{\text{idp}}^{\epsilon_p,\epsilon_s}$ in the quartic potential. All four states are degenerate for the non-interacting case with the energy $E = 12.098604$ a.u.. The approximate quantum numbers are used to calculate the effective potential in Figure 5.5. The expectation value of the Coulomb potential is used for the calculation of first-order perturbation theory in Figure 5.4.

spin	parity	$\tilde{m}_{\text{rel}}$	$\langle 1/r_{\text{rel}}\rangle$ [a.u.]	$E(\gamma = 1)$ [a.u.]
singlet	even	0.2269	1.711000091	13.65192024
triplet	even	1.0090	0.966742637	13.03986337
singlet	odd	2.0136	0.761620656	12.85001951
triplet	odd	1.0745	0.947045869	13.02380571

For two independent particles with vanishing total angular momentum we find three different classes of states for which HFR can be applied. There are pairs of singlet and triplet states with even parity and pairs with odd parity.

5.3. Quartic case

Furthermore, there are states of the form $|n_1, m_1, n_1, \text{-}m_1\rangle_{\text{idp}}^{\pm 1, \mp 1}$ which give rise to pairs of mixed parity, namely singlet even-parity states corresponding to triplet odd-parity states. In the quartic confining potential we consider 110 states in total, where the highest level is the state $|3, 0, 3, 0\rangle_{\text{idp}}^{+1,+1}$ with eigenenergy $E = 36.917640$ a.u.. Within this range there are 25 even-parity pairs and twelve mixed-parity pairs. For all of those the singlet energy is higher than the corresponding triplet energy. For the 16 odd-parity pairs considered the behaviour is inverted and all triplet energies are higher than the corresponding singlet energies. Therefore HFR holds for all the even and mixed-parity pairs, while the odd-parity pairs violate the rule. A similar effect can be observed for atomic helium, which was elaborately investigated in [168] and led to the formulation of the *alternating rule*. It is an extension to HFR that accounts for states with two open shells, i.e. $m_1 \geq 1$ and $m_2 \geq 1$. For vanishing total angular momentum odd parity states necessarily fulfil this restriction. The rule states, that for unnatural parity states, corresponding to odd states for vanishing angular momentum, HFR is reversed, such that singlet states are lower in energy than the associated triplet states.

To understand this behaviour we follow the line of argumentation for the harmonic case. As an example we pick the first pair of states that shows the violation $|1, 1, 0, -1\rangle_{\text{idp}}^{-1,\pm 1}$. We compare this pair with the degenerate pair of even states $|1, 1, 0, -1\rangle_{\text{idp}}^{+1,\pm 1}$. The common eigenenergy in the non-interacting case is $E = 12.098604$ a.u., while the energies for the interacting case are given in Table 5.2. Again we start with considering the virial theorem, Equation (2.36), and the splitting of the eigenenergy in kinetic, quartic and Coulomb potential energies (see Figure 5.4 a)). In both cases the potential energies support the trend of the total energy. This is expected for the even-parity case, but surprising for the odd case. Although one pair follows the rule, while the other violates it, the effect appears to have the same origin. The analytical values for the ratio $\lambda_4 = 6$ in the non-interacting limit, compare Equation (5.2), can be confirmed in both cases and again the explanation by Slater generally holds. The investigation in the COM and relative coordinates offered an insight for the harmonic potential, so it appears to be reasonable to consider the same for the quartic potential. In the study of the *alternating rule* in atomic helium the eigenstates were also projected on a basis in COM and relative coordinates to examine the effect [168]. We neglect the term depending on the angle φ in Equation (5.28) for a moment and assume the solution

$$\frac{1}{2\pi} \exp\left[i(\tilde{m}_{\text{COM}} \varphi_{\text{COM}} + \tilde{m}_{\text{rel}} \varphi_{\text{rel}})\right]$$

Chapter 5. Hund's first multiplicity rule

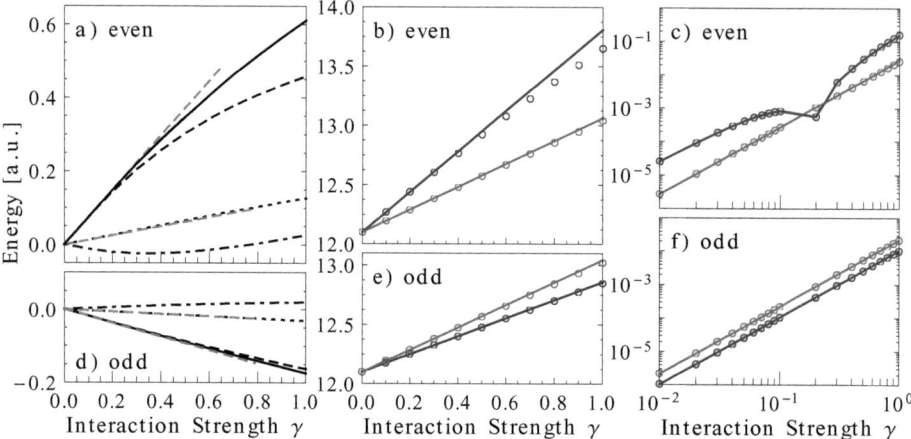

Figure 5.4: Energies of pairs of singlet and triplet states with zero angular momentum, and even $|1,1,0,-1\rangle_{\text{idp}}^{+1,\pm1}$ and odd parity $|1,1,0,-1\rangle_{\text{idp}}^{-1,\pm1}$ are compared. In a) and d) we show the differences between singlet and triplet state energy for adiabatically varying the interelectron interaction strength γ. In terms of the virial theorem (2.36) we separately consider total energy (solid line), Coulomb interaction (dashed line), quartic potential (dotted line) and kinetic energy (dot-dashed line). The dashed green lines are the linear fits in the non-interacting limit for the Coulomb interaction and the quartic confinement. The ratio $\lambda_4 \approx 6.00001$ (5.99995) of the two slopes obtained for $\gamma = 10^{-7}$ in the even (odd) parity case confirms the analytical result (5.2). The lowest states follow the *alternating rule*, the even singlet state (blue) has higher energy than the even triplet state (red) while the odd singlet state (blue) has lower energy than the odd triplet state (red). The Coulomb potential follows the same trend as the total energy. For quartic confinement the numerically exact eigenenergies E_{num} (circles) and results from first-order perturbation theory E^1 (lines) are shown in b) and e) for even and odd parity states, respectively. The deviations $E^1 - E_{\text{num}}$ are shown in c) and f), for even and odd parity states, respectively. For the singlet even case, the blue line is a guide to the eye, while for the other cases the lines are the quadratic fit. For these cases the behaviour of the states confined by a quartic potential can be understood considering first-order perturbation theory. The singlet even state, like in the harmonic potential, is significantly changed by the interaction potential.

5.3. Quartic case

for the angular part in the COM and the relative motion with the approximate not necessarily integer-valued quantum numbers $\tilde{m}_{\text{COM}}$ and $\tilde{m}_{\text{rel}}$, see Section 2.3.4. The approximate effective potential including the dependence on the angle φ reads

$$V_{\text{eff}}(r_{\text{COM}}, r_{\text{rel}}, \varphi) = -\frac{1/4 - \tilde{m}_{\text{COM}}^2}{4r_{\text{COM}}^2} - \frac{1/4 - \tilde{m}_{\text{rel}}^2}{r_{\text{rel}}^2}$$
$$+ \frac{\gamma}{r_{\text{rel}}} + 2r_{\text{COM}}^4 + \frac{1}{8}r_{\text{rel}}^4 + r_{\text{COM}}^2 r_{\text{rel}}^2 (3 - 2\sin^2\varphi). \quad (5.29)$$

In Figure 5.5 we show the limiting cases for $\sin^2\varphi = 0$ (solid line) and $\sin^2\varphi = 1$ (dashed line). The numerically obtained wave functions fit very well into these approximate effective potentials. The high angular excitation for the odd singlet state explains the violation of HFR. The angular momentum barrier shapes the localisation of the state such that it has only weak overlap with the Coulomb potential, which becomes obvious when considering the projection of the wave function on the relative radial coordinate alone, see Figure 5.5 f) (blue line).

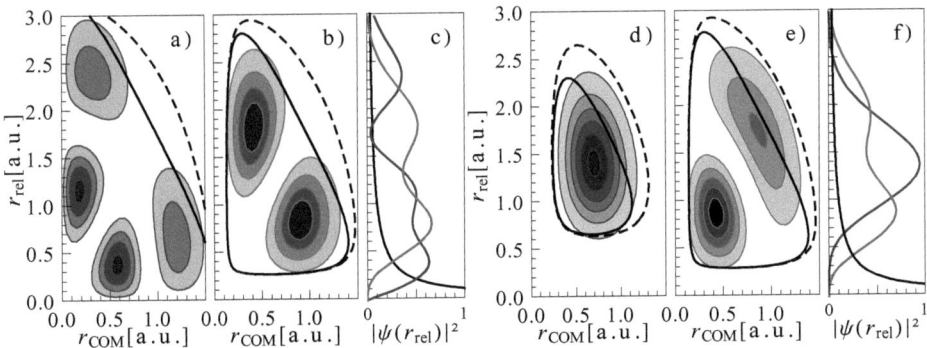

Figure 5.5: The probability densities of the states $|1, 1, 0, -1\rangle_{\text{idp}}^{\pm 1, \pm 1}$ in COM and relative radial distance for the non-interacting case ($\gamma = 0$) confined by a quartic potential. States with even parity are considered in a) singlet, b) triplet and c) the projection (integration over r_{COM}) on the relative distance, while states with odd parity are considered in d), e) and f), equivalently. The thick lines represent the equipotential lines equal to the eigenenergy of the effective potential (solid line: $\sin^2\varphi = 0$; dashed line: $\sin^2\varphi = 1$) with the numerically obtained approximate angular quantum numbers $\tilde{m}_c$ and $\tilde{m}_r$. The projections on the relative distance, c) and f), show the different effect of the Coulomb potential (schematic: solid line) on the respective singlet (blue) and triplet (red) states. The Coulomb potential has a larger effect on the singlet even than on the triplet even state, on the other hand it has a larger effect on the triplet odd state, than on the singlet odd state. This is a consequence of the different approximate relative quantum numbers shown in Table 5.3.

Chapter 5. Hund's first multiplicity rule

Finally, we also consider first-order perturbation theory for the quartic confinement. The expectation values of the Coulomb potential for the non-interacting eigenstates can be calculated numerically (see Table 5.2). The energies calculated by first-order perturbation theory are in good agreement with the numerically exact results, see Figure 5.4 b) and e). The deviation is quadratic in γ for most cases, see Figure 5.4 c) and f), except for the singlet state with even parity. Here the deviation is not quadratic, which means that the localisation of the wave function is significantly changed by the interparticle interaction.

The *alternating rule* is applicable for all singlet-triplet pairs of states with vanishing angular momentum and is valid in an energy regime up to approximately $E = 32$ a.u. (see Table 5.3). For planar confinement the even- and odd-parity states for non-vanishing angular momentum correspond to negative and positive angular momenta and are degenerate. It is not meaningful to apply the *alternating rule* for these cases.

5.4 Further confinements

Since the harmonic confinement shows a very particular behaviour with regard to the applicability of HFR, we wish to address different models that are suited to account for experimental settings. The first choice is our full model potential, which is similar to the purely quartic confinement, but more relevant from the experimental point of view. In order to extend our analysis to a broader footing we additionally investigate the hard-wall potential case. Both examples are planar models and therefore it is only reasonable to address states with vanishing angular momentum considering the *alternating rule*.

5.4.1 Full potential case

We present numerical results for the full potential

$$H_{\text{full}} = \sum_{j=1}^{2} \left[-\frac{1}{2}\nabla_j^2 + \frac{1}{2}\omega_0^2 r_j^2 + \kappa\, r_j^4 \right] + \frac{1}{r_{12}}, \qquad (5.30)$$

where we set $\omega_0 = 1$ and $\kappa = 0.1$, in the Appendix A.3 in Table A.3. For all cases the singlet state has higher energy than the corresponding triplet state, except for pairs of odd parity. We can thus conclude, that the *alternating rule* is followed by all presented states in the full potential.

5.4.2 Planar billiard

As a limiting case for confinement with positive-power-law potentials we consider the planar hard-wall potential. The Hamiltonian in modified atomic units reads

$$H_{\text{full}} = \sum_{j=1}^{2}\left[-\frac{1}{2}\nabla_j^2 + V_{\text{billiard}}(\boldsymbol{r}_j)\right] + \frac{\gamma}{r_{12}}, \qquad (5.31)$$

with

$$V_{\text{billiard}}(\boldsymbol{r}) = \begin{cases} 0 & |\boldsymbol{r}| < 1, \\ \infty & |\boldsymbol{r}| \geq 1. \end{cases} \qquad (5.32)$$

The radial solutions of the one-particle system are Bessel functions of the first kind,

$$\psi_{n,m}(r) = \mathcal{N}_{n,m}\, J_m\left(z(m, n+1)\, r\right), \qquad (5.33)$$

where $z(m,n)$ is the nth zero of $J_m(r)$ and $\mathcal{N}_{n,m}$ is a properly defined normalisation factor. The corresponding eigenenergies are given by $E_{n,m} = \frac{1}{2}[z(m,n+1)]^2$. The matrix elements

$$_{\text{idp}}\langle|n_1, m_1, n_2, m_2|\frac{1}{|\boldsymbol{r}_1 - \boldsymbol{r}_2|}|n_1', m_1', n_2', m_2'\rangle_{\text{idp}} \qquad (5.34)$$

can be calculated via the multipole expansion of the Coulomb term. For each matrix element we calculate approximately 40 terms of the multipole expansion and estimate the remainder as described in Reference [81]. With the equivalently symmetrised basis, as for our general model, Equation (1.18), we evaluate the first and second order corrections. We consider all states for $m = 0$ from the ground state up to the quadruplet of states with degenerate unperturbed eigenenergy equal to $E = 102.935354$ a.u. (see Appendix A.3, Table A.4). We compare the energies for full Coulomb interaction ($\gamma = 1$): For mixed- and even-parity pairs singlet states are higher in energy than the corresponding triplet states, while for odd-parity pairs this behaviour is reversed. Again, we can conclude, that all considered states follow the *alternating rule*.

5.5 Summary

Recent investigations [24, 106] consider the validity of the extension of HFR to excited states in a planar two-electron quantum dot model with harmonic confinement. For the application of HFR it is a necessary condition, that the singlet and triplet states compared arise from symmetrisation of the same one-electron orbitals. Our analysis shows, that in contrast to assumptions in [24, 106], HFR is in general not applicable.

We have shown the concept of fermionic entanglement witnesses to be a proper tool to investigate the applicability of HFR. The entanglement witnesses were calculated analytically in the non-interacting system by choosing an appropriate basis. Only four states, arising from the degenerate levels with $E = 3$ for the non-interacting case can be compared at all. Analysing these states, we found that the original explanation by Slater generally holds for these states. That is, the difference in the Coulombic interparticle interaction term is the dominant contribution of the energy difference between singlet and triplet states. Furthermore, the angular momentum quantum number in relative coordinates is found to be the origin of the Fermi hole and has provided us with a deeper understanding of the difference in the interaction energy.

At first sight, the behaviour for the quartic confinement potential appeared to be peculiar, since HFR is reversed for pairs of odd-parity states with zero angular momentum. Nevertheless it turns out that this behaviour is the expected one, which was already well known for atomic systems and formulated in terms of the *alternating rule* [168]. For the lowest quadruplet of degenerate states in all four symmetry classes we have shown approximate quantum numbers in COM and relative coordinates to be a meaningful concept in order to understand the origin of the *alternating rule* in the quartic confinement potential. The localisation of the wave functions is in agreement with the shape of the effective potentials according to the approximate quantum numbers. This is again the origin of the difference in the Coulomb interaction energy, which follows the same trend as the total energy in all considered cases.

Additionally, we have shown evidence that for small deviations from the harmonic confinement, which is a realistic assumption for the description of quantum dots, the *alternating rule* is also valid.

First-order perturbation theory proved to give meaningful results for the harmonic and the quartic confining potentials considered in this work. Our results for the billiard case resemble those for the quartic confinement and the *alternating rule* is again valid, up to second-order perturbation theory.

Table 5.3: Numerically obtained properties of some degenerate even and odd parity pairs in the quartic potential. For the quartic confinement all states considered with vanishing angular momentum follow the *alternating rule*. We show six examples of states, which are degenerate for the non-interacting case and split into four states, one in each symmetry class, when turning on the interaction. The behaviour of the energies is closely related to the approximate angular momentum quantum number in relative coordinates $\tilde{m}_{\text{rel}}$.

$\epsilon_p, \epsilon_s \rightarrow$	$\pm 1, \pm 1$		$+1, +1$		$+1, -1$		$-1, -1$		$-1, +1$	
	$\gamma = 0$				$\gamma = 1$					
$\|n_1, m_1, n_2, m_2\rangle_{\text{idp}}^{\epsilon_p, \epsilon_s}$	E [a.u.]		E [a.u.]	$\tilde{m}_{\text{rel}}$	E [a.u.]	$\tilde{m}_{\text{rel}}$	E [a.u.]	$\tilde{m}_{\text{rel}}$	E [a.u.]	$\tilde{m}_{\text{rel}}$
$\|1,1,0,-1\rangle_{\text{idp}}^{\epsilon_p, \epsilon_s}$	12.098 604		13.651 920	0.227	13.039 863	1.009	13.023 806	1.074	12.850 020	2.014
$\|1,2,0,-2\rangle_{\text{idp}}^{\epsilon_p, \epsilon_s}$	17.159 089		18.380 186	1.062	17.964 433	2.015	18.070 037	1.062	17.909 929	2.015
$\|2,1,0,-1\rangle_{\text{idp}}^{\epsilon_p, \epsilon_s}$	18.375 959		19.820 331	1.047	19.352 300	1.069	19.212 249	2.071	19.140 113	2.104
$\|1,3,0,-3\rangle_{\text{idp}}^{\epsilon_p, \epsilon_s}$	22.599 343		23.734 568	1.284	23.359 149	2.364	23.487 848	1.284	23.325 314	2.364
$\|2,1,1,-1\rangle_{\text{idp}}^{\epsilon_p, \epsilon_s}$	23.678 262		24.938 356	1.028	24.515 878	1.982	24.524 989	2.042	24.331 156	3.576
$\|2,2,0,-2\rangle_{\text{idp}}^{\epsilon_p, \epsilon_s}$	23.869 758		25.105 915	2.061	24.699 766	2.088	24.717 959	2.061	24.634 682	2.088

Conclusion and Perspectives

Conclusion and Perspectives

Conclusion

In this book a variety of aspects of planar two-electron quantum dots is presented. For the well-known Hooke's atom with harmonic confinement aspects of entanglement are addressed and assist to understand the applicability of Hund's first rule. Applicability, validity and origin of Hund's first rule are investigated for Hooke's atom and for further confinement potentials. We have presented a detailed description of a numerically exact treatment of planar two-electron quantum dots, which extends the common harmonic model by introducing a quartic potential term. Our approach provides an accurate characterisation of the spectrum of this system for a wide range of different parameters and symmetry classes. This has been exploited for studying anharmonic effects in the complexity of this system. Methods for the numerical time propagation are introduced and further aspects of the complexity of the system are presented. In the following we will draw the main conclusions of these investigations.

The extraordinary property of the harmonic confinement potential to be separable in centre-of-mass and relative coordinates for any kind of interaction between the particles makes this model so simple and successful in the description of quantum dots. We analysed the specific algebraic properties considering angular momentum, singlet and triplet spin states and parity symmetry. Based on these properties we present the concept of the *offset entanglement* in the non-interacting limit.

Entanglement witnesses in the non-interacting limit have revealed that the effect of the interaction is twofold. By the choice of a particular symmetry class the interaction induces entanglement for various states, already in the limit of vanishing interaction strength. Secondly, turning on the interaction further entangles the states as described in recent investigations [46, 47, 42]. The offset entanglement can be calculated for the non-interacting model solely by imposing the correct symmetrisation of the state. As a consequence it can at least qualitatively explain several effects reported in the recent literature:

(i) the non-vanishing entanglement in the non-interacting limit in harmonically confined two-electron quantum dot structures [47]; (ii) the different saturation limit for the entanglement in these models in comparison with atomic helium [47, 45]; (iii) the minimum in the linear entropy linked with the transition from cylindrical to spherical symmetry in the model involving a magnetic field [42]. Furthermore, we use the offset entanglement to determine for which states Hund's first rule is applicable in Hooke's atom. We find that Hund's first rule can only be applied to one particular pair of singlet and triplet states, including both parities. For these states we show the angular momentum quantum numbers in centre-of-mass and relative coordinates to describe the localisation of the states. The influence of the localisation on the effect of the interparticle interaction is the origin of Hund's first rule.

The harmonic confinement seems to be the exception, where Hund's first rule is applicable for a minority of states only, while in other systems, lacking the separability in centre-of-mass and relative coordinates, the applicability of Hund's first rule is generally given. The *alternating rule* is found to be valid for purely quartic confinement: Hund's first rule is reversed for odd parity states. The localisation of the states according to approximate angular momentum quantum numbers in centre-of-mass and relative coordinates is responsible for this effect. In a hard-wall potential model the *alternating rule* is also found to be valid. Thus, we believe that the *alternating rule* is valid for a wide range of confining potentials, as long as the correlation effects are to a certain extent weak.

The shell filling for few-electron quantum dots with a quartic potential perturbing the harmonic confinement has been considered and Hund's rules were found to be valid [21]. This is true for the ground states of quantum dots with a varying number of electrons. Nevertheless, in the two-electron case, the extended application of Hund's first rule to excited states holds only for even parity, while for odd parity the more general *alternating rule* needs to be applied and we have shown strong evidence that it is extensively valid. This system is fundamentally different from Hooke's atom since the separation in centre-of-mass and relative coordinates is not given. As a consequence all degrees of freedom couple and the dynamics is rendered mixed regular-chaotic. The interplay between the Coulomb interaction and the anharmonic term is responsible for a significant reduction of the regions of regular classical motion of the planar quantum dot, which otherwise is integrable. The analysis of the classical dynamics by means of the dominant frequencies of the trajectories show a mixed regular-chaotic phase space. On the quantum level, a detailed analysis of different measures of complexity supports this result. The complexity arises for small values of the quartic potential strength κ in the interacting harmonic quantum dot. The nearest-neighbour spacing distribution can be described by distributions for mixed regular-chaotic dynamics [61, 60, 63]. The occurrence of avoided crossings is typical for chaotic dynamics, while the great number of solitonic states characterise a mixed regime. The information entropy of the basis representation in the non-interacting harmonic basis revealed a rising complexity of the states for higher energies. Finally a dynamic property, the quantum fidelity susceptibility of eigenstates, particularly indicates the loss of separability. We confirm the behaviour of the susceptibility by numerically propagating the equations of motion and by an analytical derivation based solely on static properties of the system.

Depending on the material of the quantum dot, magnetic fields of the order of some Tesla might suppress deviations from the harmonic potential. Increasing the magnetic field breaks the condition $\omega_{xy} \ll \omega_z$ justifying the two-dimensional approximation. Therefore, the magnetic field can be used for a controlled transition from an effectively two-dimensional to a three-dimensional system. The effect of the third dimension can be taken into account by the choice of the parameters [101]: The interaction strength γ decreases towards the three-dimensional system or equivalently the confining potential strengths ω and κ increase, as a consequence of our choice of units. Since our results do not strongly depend on these parameters, our analysis of the complexity is to a large extent independent of the dimensionality.

The investigations of the quantum fidelity of eigenstates are at the same time the first step towards the investigation of a certain phenomenon reported for many-body systems. The quantum fidelity decay for such systems described with mean-field approaches show an unusual decay behaviour [54, 55, 56]. Our model of a planar two-electron quantum dot with harmonic confinement and a quartic perturbation that renders the dynamics mixed regular-chaotic can be used address this phenomenon. It needs to be clarified, if the effect is due to the mean-field treatment or associated to the complex dynamics. For this purpose we have implemented, tested and further developed different time-evolution methods of which we discuss three in particular. The Fatunla method has the advantage of an adaptive time step and is particularly suited for time-depending problems. Short computation times and a very good conservation of the norm feature the Arnoldi method. A very accurate and sophisticated method is the iterative predictor-corrector scheme introduced. In the latter, we present a method to adapt the time step based on the precision of the norm, which is in

Conclusion and Perspectives

general close to machine precision.

Based on a versatile numerical approach we perform a detailed characterisation of the complexity of the anharmonic model. We derive general results for the entanglement of models with harmonic confinement and arbitrary interaction. A profound investigation of Hund's first rule for quantum dots with different confinements is presented. Furthermore, a basis for exact investigations of time-dependent phenomena is provided and first results are presented. Eventually, we shed some light on a variety of aspects of two-electron quantum dots.

Perspectives

One of the main results of this book is an exact numerical approach to a planar two-electron quantum dot with complex dynamics including accurate time-evolution methods. Beyond the investigations of the quantum fidelity of eigenstates different initial states, in particular Gaussian distributions, can be used. It might be possible to confirm the unusual behaviour of the fidelity decay reported for many-body systems. On the other hand, the manipulation of the perturbing potential in order to fulfil the condition for quantum freezing might unmask the unusual decay to be caused by the mean-field approximation.

The algebraic representation of observables is easily evaluated and in parts already implemented in the numerical approach. In particular the dipole operator, implemented for the calculation of the oscillator strengths, can be used to calculate the interaction of the quantum dot with electromagnetic fields in dipole approximation. This offers the opportunity to investigate phenomena connected to the interactions of quantum dots with laser pulses.

Another system which is easily accessible with our method is obtained by changing the sign in front of the harmonic potential. This minor change will lead to the description of a planar two-electron quantum ring [186].

For systems showing a particular separation of variables further investigations of the offset entanglement might clarify properties of the interacting model. For example, in the case of cylindrically symmetric quantum dots with applied magnetic field reported in [42], we expect the entanglement of excited states in the non-interacting limit to depend discontinuously on the magnetic field.

Appendix A

Addenda to the book

A.1 Modified atomic units

We apply modified atomic units ($\gamma \equiv 1$), where we set $\hbar = m_* = (4\pi\epsilon_0\epsilon)^{-1} = q_e = k_B = 1$. The natural scales are the modified Bohr-radius

$$a_0 = \frac{4\pi\epsilon_0\epsilon\hbar^2}{m_* q_e^2},$$

the modified Hartree energy

$$E_{\rm h} = \frac{m_* q_e^4}{(4\pi\epsilon_0\epsilon\hbar)^2},$$

the natural frequency

$$\nu_0 = \frac{E_{\rm h}}{\hbar} = \frac{m_* q_e^4}{\hbar^3 (4\pi\epsilon_0\epsilon)^2},$$

and the natural timescale

$$t_0 = \frac{\hbar}{E_{\rm h}} = \frac{\hbar^3 (4\pi\epsilon_0\epsilon)^2}{m_* q_e^4}.$$

A typical value for the confining potential is $E_0 = \hbar\tilde{\omega}_0 = 3\,{\rm meV}$ [20], which

Table A.1: Solid state quantum dots are most commonly realised on substrates of Indium Arsenide (InAs) and Gallium Arsenide (GaAs). For both cases we supply the values of the effective electron mass m^*, the dielectric constant ϵ and the natural scales.

Substrate	m^* [m_e]	ϵ [ϵ_0]	a_0 [nm]	$E_{\rm h}$ [meV]	ν_0 [THz]	t_0 [psec]
InAs	0.023	15.15	35	2.7	4.14	0.24
GaAs	0.063	12.9	11	10	15.65	0.066

implies that ω_0 is of order unity. The ratio of the harmonic confining potential to the natural energy scale of one Hartree (E_h) is expressed as $\omega_0 = E_0/E_\mathrm{h}$. We set $\omega_0 \equiv 1$ in most of our calculations.

The effect of a magnetic field depends strongly on the material. A magnetic field of one Tesla leads to $\omega_B \approx 1$ for InAs, while for GaAs it leads to $\omega_B \approx 0.1$.

A.2   Parameters of the predictor-corrector scheme

Implicit Radau-type four-step Runge-Kutta method

The corrector for the predictor-corrector scheme discussed in Section 4.2.4 is based on an implicit Radau II A type Runge-Kutta method of order 7. For this method the consecutive vectors are calculated by

$$\mathbf{y}_{n+1} = \mathbf{y}_n - \mathrm{i}\delta t \sum_{i=1}^{4} b_i \mathbf{H}(t_n + \tau_i) \mathbf{Y}_i.$$

The intermediate vectors $\mathbf{Y}_i$ are calculated with the following implicit scheme

$$\begin{pmatrix} \mathbf{Y}_1 \\ \vdots \\ \mathbf{Y}_4 \end{pmatrix} = \begin{pmatrix} \mathbf{y}_n \\ \vdots \\ \mathbf{y}_n \end{pmatrix} - \mathrm{i}\delta t \begin{pmatrix} a_{11}\mathbf{H}(t_n + \tau_1) & \cdots & a_{14}\mathbf{H}(t_n + \tau_4) \\ \vdots & \ddots & \vdots \\ a_{41}\mathbf{H}(t_n + \tau_1) & \cdots & a_{44}\mathbf{H}(t_n + \tau_4) \end{pmatrix} \begin{pmatrix} \mathbf{Y}_1 \\ \vdots \\ \mathbf{Y}_4 \end{pmatrix},$$

where $\boldsymbol{a} = (a_{ij})$ and $\boldsymbol{b} = (b_i)$ are the parameters of the Runge-Kutta-method. The intermediate time steps are related to the parameters $c_i = \tau_i/\delta t$. All parameters can in general derived by a collocation method with arbitrary precision [123]. We show the parameters in the well-known Butcher tableau with only six digits (see Table A.2).

Table A.2: Butcher tableau for the implicit Radau II A type Runge-Kutta method. The first column contains the parameters c_i counting downwards. The last row contains the parameters b_j counting from left to right. The matrix in the middle contains the parameters a_{ij}, accordingly. Notice that $\sum_{j=1}^{4} a_{ij} = c_i$ as usual for Runge-Kutta methods. The method fulfils the property $b_j = a_{4j}$ and is thus called to be "stiffly accurate" and is particularly suited for stiff problems.

0.088588	0.112999	−0.040309	0.025802	−0.009905
0.409467	0.234384	0.206893	−0.047857	0.016047
0.787659	0.216682	0.406123	0.189036	−0.024182
1.000000	0.220462	0.388193	0.328844	0.062500
$i \downarrow j \rightarrow$	0.220462	0.388193	0.328844	0.062500

Parameters of the iterative corrector scheme

The diagonal matrix $\mathbf{d} = \mathrm{diag}(d_{11},..,d_{44})$ chosen for fast convergence of the iterative corrector scheme is taken from [187]:

$$d_{11} = 0.32049937,$$
$$d_{22} = 0.08915379,$$
$$d_{33} = 0.18173957,$$
$$d_{44} = 0.23336280.$$

A.3 Spectral properties for the full potential and the planar billiard

This addendum complements Chapter 5. Spectral properties of the full potential ($\omega_0 = 1$, $\kappa = 0.1$, and $\gamma = 1$) and the planar two-electron billiard are presented in Tables A.3 and A.4, respectively.

Table A.3: For the full potential we show all states ($|n_1, m_1, n_2, m_2\rangle_{\mathrm{idp}}^{\epsilon_p,\epsilon_s}$) with vanishing angular momentum up to the fourth state that exists only as singlet even $|3,0,3,0\rangle_{\mathrm{idp}}^{+1,+1}$. All pairs of singlet and triplet spin symmetry presented here follow the *alternating rule*. For the parity mixing pairs the singlet has higher eigenenergy than the triplet in all cases.

$\epsilon_p, \epsilon_s \rightarrow$	$\pm 1, \pm 1$	$+1, +1$	$+1, -1$	$-1, -1$	$-1, +1$
States	$E\,(\gamma = 0)$	\multicolumn{4}{c}{$E\,(\gamma = 1)$}			
$\|0,0,0,0\rangle_{\mathrm{idp}}^{\epsilon_p,\epsilon_s}$	2.300376	3.445300			
$\|0,1,0,-1\rangle_{\mathrm{idp}}^{\epsilon_p,\epsilon_s}$	4.828681	5.901348		5.526039	
$\|1,0,0,0\rangle_{\mathrm{idp}}^{\epsilon_p,\epsilon_s}$	5.026830	6.210941	5.745411		
$\|0,2,0,-2\rangle_{\mathrm{idp}}^{\epsilon_p,\epsilon_s}$	7.544645	8.304078		8.221913	
$\|1,0,1,0\rangle_{\mathrm{idp}}^{\epsilon_p,\epsilon_s}$	7.753283	8.658231			
$\|1,1,0,-1\rangle_{\mathrm{idp}}^{\epsilon_p,\epsilon_s}$	7.809767	8.939785	8.507135	8.509367	8.374386
$\|2,0,0,0\rangle_{\mathrm{idp}}^{\epsilon_p,\epsilon_s}$	8.205149	9.297429	8.932604		
$\|0,3,0,-3\rangle_{\mathrm{idp}}^{\epsilon_p,\epsilon_s}$	10.422142	11.168524		11.067339	
$\|1,2,0,-2\rangle_{\mathrm{idp}}^{\epsilon_p,\epsilon_s}$	10.743092	11.550716	11.311839	11.346638	11.298899
$\|1,1,1,-1\rangle_{\mathrm{idp}}^{\epsilon_p,\epsilon_s}$	10.790854	11.660480		11.437097	
$\|2,0,1,0\rangle_{\mathrm{idp}}^{\epsilon_p,\epsilon_s}$	10.931602	11.915729	11.544597		
$\|2,1,0,-1\rangle_{\mathrm{idp}}^{\epsilon_p,\epsilon_s}$	11.169260	12.255326	11.884325	11.804700	11.736365

Continued on next page

Chapter A. Addenda to the book

Table A.3 – continued from previous page

$\epsilon_p, \epsilon_s \rightarrow$	$\pm 1, \pm 1$	$+1, +1$	$+1, -1$	$-1, -1$	$-1, +1$
States	$E(\gamma = 0)$	\multicolumn{4}{c}{$E(\gamma = 1)$}			
$\|3,0,0,0\rangle_{\text{idp}}^{\epsilon_p, \epsilon_s}$	11.718736	12.694313	12.429348		
$\|0,4,0,-4\rangle_{\text{idp}}^{\epsilon_p, \epsilon_s}$	13.442625	14.161967		14.063219	
$\|1,3,0,-3\rangle_{\text{idp}}^{\epsilon_p, \epsilon_s}$	13.811615	14.545183	14.356594	14.452876	14.341369
$\|1,2,1,-2\rangle_{\text{idp}}^{\epsilon_p, \epsilon_s}$	13.941538	14.605366		14.518543	
$\|2,0,2,0\rangle_{\text{idp}}^{\epsilon_p, \epsilon_s}$	14.109921	14.864683			
$\|2,1,1,-1\rangle_{\text{idp}}^{\epsilon_p, \epsilon_s}$	14.150347	15.029219	14.762885	14.756618	14.625066
$\|2,2,0,-2\rangle_{\text{idp}}^{\epsilon_p, \epsilon_s}$	14.268852	15.116291	14.849053	14.906673	14.830529
$\|3,0,1,0\rangle_{\text{idp}}^{\epsilon_p, \epsilon_s}$	14.445189	15.433439	15.088643		
$\|3,1,0,-1\rangle_{\text{idp}}^{\epsilon_p, \epsilon_s}$	14.828671	15.814453	15.532502	15.412681	15.379656
$\|4,0,0,0\rangle_{\text{idp}}^{\epsilon_p, \epsilon_s}$	15.507363	16.388931	16.196457		
$\|0,5,0,-5\rangle_{\text{idp}}^{\epsilon_p, \epsilon_s}$	16.592132	17.286283		17.192212	
$\|1,4,0,-4\rangle_{\text{idp}}^{\epsilon_p, \epsilon_s}$	17.003417	17.738025	17.524732	17.626659	17.512953
$\|1,3,1,-3\rangle_{\text{idp}}^{\epsilon_p, \epsilon_s}$	17.201089	17.825499		17.750490	
$\|2,2,1,-2\rangle_{\text{idp}}^{\epsilon_p, \epsilon_s}$	17.467298	18.143407	17.959234	17.993150	17.948675
$\|2,3,0,-3\rangle_{\text{idp}}^{\epsilon_p, \epsilon_s}$	17.490745	18.232610	18.045281	18.060870	18.028437
$\|2,1,2,-1\rangle_{\text{idp}}^{\epsilon_p, \epsilon_s}$	17.509840	18.262858		18.125871	
$\|3,0,2,0\rangle_{\text{idp}}^{\epsilon_p, \epsilon_s}$	17.623508	18.476730	18.175475		
$\|3,1,1,-1\rangle_{\text{idp}}^{\epsilon_p, \epsilon_s}$	17.809757	18.752802	18.458853	18.386646	18.304178
$\|3,2,0,-2\rangle_{\text{idp}}^{\epsilon_p, \epsilon_s}$	18.065858	18.820033	18.635417	18.650041	18.613835
$\|4,0,1,0\rangle_{\text{idp}}^{\epsilon_p, \epsilon_s}$	18.233816	19.157663	18.875963		
$\|4,1,0,-1\rangle_{\text{idp}}^{\epsilon_p, \epsilon_s}$	18.741379	19.630870	19.426534	19.291770	19.274082
$\|5,0,0,0\rangle_{\text{idp}}^{\epsilon_p, \epsilon_s}$	19.532674	20.344307	20.200976		
$\|0,6,0,-6\rangle_{\text{idp}}^{\epsilon_p, \epsilon_s}$	19.859704	20.531683		20.442313	
$\|1,5,0,-5\rangle_{\text{idp}}^{\epsilon_p, \epsilon_s}$	20.308870	21.017504	20.810666	20.912889	20.801416
$\|1,4,1,-4\rangle_{\text{idp}}^{\epsilon_p, \epsilon_s}$	20.564210	21.164293		21.094610	
$\|3,3,0,-3\rangle_{\text{idp}}^{\epsilon_p, \epsilon_s}$	20.824908	21.499174	21.341913	21.393522	21.329198
$\|2,3,1,-3\rangle_{\text{idp}}^{\epsilon_p, \epsilon_s}$	20.880218	21.523390	21.375837	21.464398	21.358846
$\|2,2,2,-2\rangle_{\text{idp}}^{\epsilon_p, \epsilon_s}$	20.993059	21.602197		21.517722	
$\|3,0,3,0\rangle_{\text{idp}}^{\epsilon_p, \epsilon_s}$	21.137096	21.796490			

A.3. Spectral properties for the full potential and the planar billiard

Table A.4: For the billiard potential V_{billiard} we show unperturbed energies and the first- and second-order corrections for the first 67 states in all four symmetry classes with vanishing angular momentum. Values are given in modified atomic units. All presented states follow the *alternating rule* considering full Coulomb interaction ($\gamma = 1$).

$\epsilon_p, \epsilon_s \to$	$\pm 1, \pm 1$	$+1, +1$		$+1, -1$		$-1, -1$		$-1, +1$	
$\lvert n_1, m_1, n_2, m_2 \rangle_{\text{idp}}^{\epsilon_p, \epsilon_s}$	$E(\gamma = 0)$	$\mathcal{O}(\gamma)$	$\mathcal{O}(\gamma^2)$	$\mathcal{O}(\gamma)$	$\mathcal{O}(\gamma^2)$	$\mathcal{O}(\gamma)$	$\mathcal{O}(\gamma^2)$	$\mathcal{O}(\gamma)$	$\mathcal{O}(\gamma^2)$
$\lvert 0,0,0,0 \rangle_{\text{idp}}^{\epsilon_p,\epsilon_s}$	5.783186	2.596	−0.96						
$\lvert 0,1,0,-1 \rangle_{\text{idp}}^{\epsilon_p,\epsilon_s}$	14.681971	3.322	−0.70			0.903	−0.08		
$\lvert 1,0,0,0 \rangle_{\text{idp}}^{\epsilon_p,\epsilon_s}$	18.127224	3.120	−0.61	1.660					
$\lvert 0,2,0,-2 \rangle_{\text{idp}}^{\epsilon_p,\epsilon_s}$	26.374616	2.747	−0.26			1.207	−0.05		
$\lvert 1,0,1,0 \rangle_{\text{idp}}^{\epsilon_p,\epsilon_s}$	30.471262	2.481	−0.35						
$\lvert 1,1,0,-1 \rangle_{\text{idp}}^{\epsilon_p,\epsilon_s}$	31.950213	3.921	−0.49	1.908	−0.07	1.908	−0.09	1.178	−0.03
$\lvert 2,0,0,0 \rangle_{\text{idp}}^{\epsilon_p,\epsilon_s}$	40.335096	2.815	−0.36	1.923	−0.08				
$\lvert 0,3,0,-3 \rangle_{\text{idp}}^{\epsilon_p,\epsilon_s}$	40.706466	2.510	0.05			1.323	−0.04		
$\lvert 1,2,0,-2 \rangle_{\text{idp}}^{\epsilon_p,\epsilon_s}$	48.612308	3.136	−0.27	1.636	−0.06	1.636	−0.04	1.351	−0.03
$\lvert 1,1,1,-1 \rangle_{\text{idp}}^{\epsilon_p,\epsilon_s}$	49.218456	3.279	−0.40			1.134	−0.03		
$\lvert 2,0,1,0 \rangle_{\text{idp}}^{\epsilon_p,\epsilon_s}$	52.679135	3.092	−0.08	1.658	0.00				
$\lvert 0,4,0,-4 \rangle_{\text{idp}}^{\epsilon_p,\epsilon_s}$	57.582941	2.380	0.18			1.388	−0.02		
$\lvert 2,1,0,-1 \rangle_{\text{idp}}^{\epsilon_p,\epsilon_s}$	59.090712	3.533	−0.29	2.193	−0.07	2.193	−0.09	1.250	−0.03
$\lvert 1,3,0,-3 \rangle_{\text{idp}}^{\epsilon_p,\epsilon_s}$	67.992019	2.797	0.12	1.539	−0.01	1.539	−0.02	1.395	−0.02
$\lvert 1,2,1,-2 \rangle_{\text{idp}}^{\epsilon_p,\epsilon_s}$	70.849999	2.743	−0.75			1.427	−0.03		
$\lvert 3,0,0,0 \rangle_{\text{idp}}^{\epsilon_p,\epsilon_s}$	72.411735	2.693	0.06	2.032	−0.08				
$\lvert 2,0,2,0 \rangle_{\text{idp}}^{\epsilon_p,\epsilon_s}$	74.887007	2.439	−0.28						
$\lvert 2,1,1,-1 \rangle_{\text{idp}}^{\epsilon_p,\epsilon_s}$	76.358955	4.084	−0.34	2.001	−0.02	2.001	−0.03	1.238	−0.01

Continued on next page

Table A.4 – continued from previous page

$\epsilon_p, \epsilon_s \to$	$\pm 1, \pm 1$	$+1, +1$		$+1, -1$		$-1, -1$		$-1, +1$	
$\|n_1, m_1, n_2, m_2\rangle_{\text{idp}}^{\epsilon_p, \epsilon_s}$	$E(\gamma = 0)$	$\mathcal{O}(\gamma)$	$\mathcal{O}(\gamma^2)$	$\mathcal{O}(\gamma)$	$\mathcal{O}(\gamma^2)$	$\mathcal{O}(\gamma)$	$\mathcal{O}(\gamma^2)$	$\mathcal{O}(\gamma)$	$\mathcal{O}(\gamma^2)$
$\|0,5,0,-5\rangle_{\text{idp}}^{\epsilon_p,\epsilon_s}$	76.938928	2.299	0.57			1.432	−0.01		
$\|2,2,0,-2\rangle_{\text{idp}}^{\epsilon_p,\epsilon_s}$	80.697663	2.833	−0.06	1.823	−0.05	1.823	−0.04	1.438	−0.03
$\|3,0,1,0\rangle_{\text{idp}}^{\epsilon_p,\epsilon_s}$	84.755773	2.812	0.10	1.901	0.01				
$\|1,4,0,-4\rangle_{\text{idp}}^{\epsilon_p,\epsilon_s}$	90.005368	2.602	0.25	1.497	0.01	1.497	0.01	1.414	−0.01
$\|1,3,1,-3\rangle_{\text{idp}}^{\epsilon_p,\epsilon_s}$	95.277573	2.509	−0.95			1.519	−0.03		
$\|3,1,0,-1\rangle_{\text{idp}}^{\epsilon_p,\epsilon_s}$	96.101369	3.363	0.59	2.328	−0.07	2.328	−0.09	1.235	−0.02
$\|0,6,0,-6\rangle_{\text{idp}}^{\epsilon_p,\epsilon_s}$	98.726272	2.243	0.21			1.466	0.01		
$\|2,2,1,-2\rangle_{\text{idp}}^{\epsilon_p,\epsilon_s}$	102.935354	3.329	−0.62	1.743	−0.07	1.743	−0.01	1.436	−0.01

A.4 Level structure of the harmonic model

The system of two non-interacting Fermions in a planar harmonic confinement is described by the Hamiltonian

$$H = -\frac{1}{2}\left(\nabla_1^2 + \nabla_2^2\right) + \frac{\omega_0^2}{2}\left(r_1^2 + r_2^2\right). \tag{A.1}$$

The energy levels are given by $E_n = \omega(n+2)$ with the principal quantum number $n \in \mathbb{N}_0$ The degeneracy of each of these levels is $\#(E_n) = \frac{1}{6}(n+1)(n+2)(n+3)$ leading to a total number of $\frac{1}{24}(N+1)(N+2)(N+3)(N+4)$ states with principal quantum number $n \leq N$. We distinguish between states with different total angular momentum $m \in \mathbb{Z}$. States can have singlet ($\epsilon_s = +1$) or triplet ($\epsilon_s = -1$) spin symmetry. For zero angular momentum the states can have even ($\epsilon_p = +1$) or odd ($\epsilon_p = -1$) parity (see Table A.5). For non-vanishing angular momentum the degeneracy for positive or negative angular momenta is equal (see Tables A.6 and A.7). Symmetrising with the parity operator for these cases gives the same degeneracy for even and odd parity as for positive and negative angular momenta.

Chapter A. Addenda to the book

Table A.5: Degeneracies for two Fermions in the non-interacting harmonic model for zero angular momentum. We show the principal quantum numbers n, the energies $E_n = 2 + n$ and the corresponding degeneracies $\#E_n$ in each of the four symmetry classes. We also show the cumulative number of states $\Sigma(\#E_n)$ in each symmetry class.

		$\epsilon_s = +1$, singlet				$\epsilon_s = -1$, triplet			
		$\epsilon_p = +1$		$\epsilon_p = -1$		$\epsilon_p = +1$		$\epsilon_p = -1$	
n	E_n	$\#E_n$	$\Sigma(\#E_n)$	$\#E_n$	$\Sigma(\#E_n)$	$\#E_n$	$\Sigma(\#E_n)$	$\#E_n$	$\Sigma(\#E_n)$
0	2	1	1	0	0	0	0	0	0
2	4	2	3	0	0	1	1	1	1
4	6	4	7	1	1	2	3	2	3
6	8	6	13	2	3	4	7	4	7
8	10	9	22	4	7	6	13	6	13
10	12	12	34	6	13	9	22	9	22
12	14	16	50	9	22	12	34	12	34
14	16	20	70	12	34	16	50	16	50
16	18	25	95	16	50	20	70	20	70
18	20	30	125	20	70	25	95	25	95
20	22	36	161	25	95	30	125	30	125
22	24	42	203	30	125	36	161	36	161
24	26	49	252	36	161	42	203	42	203
26	28	56	308	42	203	49	252	49	252
28	30	64	372	49	252	56	308	56	308
30	32	72	444	56	308	64	372	64	372
32	34	81	525	64	372	72	444	72	444
34	36	90	615	72	444	81	525	81	525
36	38	100	715	81	525	90	615	90	615
38	40	110	825	90	615	100	715	100	715
40	42	121	946	100	715	110	825	110	825
42	44	132	1078	110	825	121	946	121	946
44	46	144	1222	121	946	132	1078	132	1078
46	48	156	1378	132	1078	144	1222	144	1222
48	50	169	1547	144	1222	156	1378	156	1378
50	52	182	1729	156	1378	169	1547	169	1547
52	54	196	1925	169	1547	182	1729	182	1729
54	56	210	2135	182	1729	196	1925	196	1925
56	58	225	2360	196	1925	210	2135	210	2135
58	60	240	2600	210	2135	225	2360	225	2360

A.4. Level structure of the harmonic model

Table A.6: Degeneracies for two Fermions in the non-interacting harmonic model with odd angular momenta. We show the principal quantum numbers n, the energies $E_n = 2+n$ and the corresponding degeneracies $\#E_n$. The degeneracies are equal for positive and negative angular momentum and for singlet and triplet spin symmetry. We also show the cumulative number of states $\Sigma(\#E_n)$.

		$m = \pm 1$		$m = \pm 3$		$m = \pm 5$	
n	E_n	$\#E_n$	$\Sigma(\#E_n)$	$\#E_n$	$\Sigma(\#E_n)$	$\#E_n$	$\Sigma(\#E_n)$
1	3	1	1	0	0	0	0
3	5	3	4	2	2	0	0
5	7	6	10	5	7	3	3
7	9	10	20	9	16	7	10
9	11	15	35	14	30	12	22
11	13	21	56	20	50	18	40
13	15	28	84	27	77	25	65
15	17	36	120	35	112	33	98
17	19	45	165	44	156	42	140
19	21	55	220	54	210	52	192
21	23	66	286	65	275	63	255
23	25	78	364	77	352	75	330
25	27	91	455	90	442	88	418
27	29	105	560	104	546	102	520
29	31	120	680	119	665	117	637
31	33	136	816	135	800	133	770
33	35	153	969	152	952	150	920
35	37	171	1140	170	1122	168	1088
37	39	190	1330	189	1311	187	1275
39	41	210	1540	209	1520	207	1482
41	43	231	1771	230	1750	228	1710
43	45	253	2024	252	2002	250	1960
45	47	276	2300	275	2277	273	2233
47	49	300	2600	299	2576	297	2530
49	51	325	2925	324	2900	322	2852
51	53	351	3276	350	3250	348	3200
53	55	378	3654	377	3627	375	3575
55	57	406	4060	405	4032	403	3978
57	59	435	4495	434	4466	432	4410

Chapter A. Addenda to the book

Table A.7: Degeneracies for two Fermions in the non-interacting harmonic model for even angular momenta. We show the principal quantum numbers n, the energies $E_n = 2+n$ and the corresponding degeneracies $\#E_n$. The degeneracies are equal for positive and negative angular momenta. We also show the cumulative number of states $\Sigma(\#E_n)$.

		$m = \pm 2$				$m = \pm 4$			
		$\epsilon_s = +1$, singlet		$\epsilon_s = -1$, triplet		$\epsilon_s = +1$, singlet		$\epsilon_s = -1$, triplet	
n	E_n	$\#E_n$	$\Sigma(\#E_n)$	$\#E_n$	$\Sigma(\#E_n)$	$\#E_n$	$\Sigma(\#E_n)$	$\#E_n$	$\Sigma(\#E_n)$
2	4	2	2	1	1	0	0	0	0
4	6	4	6	4	5	3	3	2	2
6	8	8	14	7	12	6	9	6	8
8	10	12	26	12	24	11	20	10	18
10	12	18	44	17	41	16	36	16	34
12	14	24	68	24	65	23	59	22	56
14	16	32	100	31	96	30	89	30	86
16	18	40	140	40	136	39	128	38	124
18	20	50	190	49	185	48	176	48	172
20	22	60	250	60	245	59	235	58	230
22	24	72	322	71	316	70	305	70	300
24	26	84	406	84	400	83	388	82	382
26	28	98	504	97	497	96	484	96	478
28	30	112	616	112	609	111	595	110	588
30	32	128	744	127	736	126	721	126	714
32	34	144	888	144	880	143	864	142	856
34	36	162	1050	161	1041	160	1024	160	1016
36	38	180	1230	180	1221	179	1203	178	1194
38	40	200	1430	199	1420	198	1401	198	1392
40	42	220	1650	220	1640	219	1620	218	1610
42	44	242	1892	241	1881	240	1860	240	1850
44	46	264	2156	264	2145	263	2123	262	2112
46	48	288	2444	287	2432	286	2409	286	2398
48	50	312	2756	312	2744	311	2720	310	2708
50	52	338	3094	337	3081	336	3056	336	3044
52	54	364	3458	364	3445	363	3419	362	3406
54	56	392	3850	391	3836	390	3809	390	3796
56	58	420	4270	420	4256	419	4228	418	4214
58	60	450	4720	449	4705	448	4676	448	4662

List of Tables

2.1 Numerically and analytically calculated eigenenergies E and expectation values $\langle dH/d\alpha \rangle$, $\langle R_{\text{COM}} \rangle$ and $\langle r_{\text{rel}} \rangle$ for Hooke's atom . 47

2.2 Our numerical results are compared to the one-particle solutions taken from [88] for the planar quartic oscillator 48

5.1 Entanglement witnesses for low-lying states in the non-interacting harmonic model 104

5.2 Properties of the states $|1,1,0,-1\rangle_{\text{idp}}^{\epsilon_p,\epsilon_s}$ in the quartic potential . 110

5.3 Properties of some degenerate even and odd parity pairs in the quartic potential 117

A.1 Effective electron mass m^*, the dielectric constant ϵ and the natural scales for InAs and GaAs 123

A.2 Butcher tableau for the Radau II A type implicit Runge-Kutta method 124

A.3 Eigenenergies in the full potential for singlet/triplet and even/odd parity 125

A.4 Energies and perturbative corrections for states in the billiard model 127

A.5 Degeneracies for two Fermions in the non-interacting harmonic model with zero angular momentum 130

A.6 Degeneracies for two Fermions in the non-interacting harmonic model with odd angular momenta 131

A.7 Degeneracies for two Fermions in the non-interacting harmonic model for even angular momenta 132

List of Figures

1.1	Low-lying eigenenergies of the relative motion depending on γ and ω_B	27
1.2	The spectrum of all states arising from harmonic oscillator energies $E \leq 6$ for $\omega = 1$ with varying γ and κ	29
1.3	Ground state transition between singlet and triplet states for Hooke's atom with quartic perturbation and a magnetic field . .	31
2.1	Convergence properties depending on the scaling factor α	45
3.1	Dominant frequencies of the planar classical dot with $\omega_0 = 1$, $\gamma = 1$ and $\kappa = 0.1$	60
3.2	The procedure of unfolding and a typical result for the nearest-neighbour spacing distribution with appropriate fit functions . .	62
3.3	Fit parameters for the nearest-neighbour spacing distributions for variation of ω_0, κ and γ	64
3.4	Characteristics and detection of two examples of avoided crossings occurring for variation of κ	68
3.5	Characterisation of the distributions of the widths of the avoided crossings	69
3.6	Characterisation of avoided level crossings between solitonic states	70
3.7	Information entropies of states with for four combinations of the parameters κ and γ	74
4.1	Properties of the Fatunla time evolution method calculating the quantum fidelity susceptibility of an eigenstate.	84
4.2	Properties of the Arnoldi time evolution method calculating the quantum fidelity susceptibility of an eigenstate	87
4.3	Properties of the predictor-corrector scheme calculating the quantum fidelity susceptibility of an eigenstate	90
4.4	Quantum fidelity susceptibilities for four eigenstates in Hooke's atom	93

4.5	Susceptibility for eigenstates for quartic confinement	94
5.1	Linear entropies $\mathcal{E}_\mathrm{L}$ for all eigenstates in the non-interacting harmonic model up to principal quantum number $n = 13$	103
5.2	Contributions to the energy differences and results from perturbation theory between the corresponding singlet and triplet states in the harmonic model turning on the interaction	107
5.3	Probability densities of the corresponding singlet and triplet states in the harmonic model with and without interaction	109
5.4	Contributions to the energy differences and results from perturbation theory for even and odd parity pairs in the quartic oscillator turning on the interaction	112
5.5	Probability densities for even and odd parity pairs in the quartic potential	113

Bibliography

[1] A. I. Ekimov and A. A. Onushchenko. *Quantum size effect in three-dimensional microscopic semiconductor crystals.* J. Exp. Theor. Phys. Lett. **34** 363 (1981).

[2] L. E. Brus. *Electron–electron and electron-hole interactions in small semiconductor crystallites: The size dependence of the lowest excited electronic state.* J. Chem. Phys. **80** 4403 (1984).

[3] M. A. Reed, J. N. Randall, R. J. Aggarwal, R. J. Matyi, T. M. Moore and A. E. Wetsel. *Observation of discrete electronic states in a zero-dimensional semiconductor nanostructure.* Phys. Rev. Lett. **60** 535 (1988).

[4] D. Bimberg. *Der Zoo der Quantenpunkte.* Physik Journal **09/2006** 43 (2006).

[5] Y. Arakawa and H. Sakaki. *Multidimensional quantum well laser and temperature dependence of its threshold current.* Appl. Phys. Lett. **40** 939 (1982).

[6] R. Hanson, L. P. Kouwenhoven, J. R. Petta, S. Tarucha and L. M. K. Vandersypen. *Spins in few-electron quantum dots.* Rev. Mod. Phys. **79** 1217 (2007).

[7] Q. Sun, Y. A. Wang, L. S. Li, D. Wang, T. Zhu, J. Xu, C. Yang and Y. Li. *Bright, multicoloured light-emitting diodes based on quantum dots.* Nature Photon. **1** 717 (2007).

[8] P. O. Anikeeva, J. E. Halpert, M. G. Bawendi and V. Bulovic. *Quantum Dot Light-Emitting Devices with Electroluminescence Tunable over the Entire Visible Spectrum.* Nano Lett. **9** 2532 (2009).

[9] O. E. Semonin, J. M. Luther, S. Choi, H.-Y. Chen, J. Gao, A. J. Nozik and M. C. Beard. *Peak External Photocurrent Quantum Efficiency Exceeding 100MEG in a Quantum Dot Solar Cell.* Science **334** 1530 (2011).

[10] O. E. Semonin and M. C. Beard. *Quantum dots for next-generation photovoltaics*. Mater. Today **15** 508 (2012).

[11] H. S. Friedrich. *Theoretical Atomic Physics* (Springer, Berlin Heidelberg, 2005).

[12] D. Pfannkuche, V. Gudmundsson and P. A. Maksym. *Comparison of a Hartree, a Hartree-Fock, and an exact treatment of quantum-dot helium*. Phys. Rev. B **47** 2244 (1993).

[13] N. F. Johnson. *Quantum dots: few-body, low-dimensional systems*. J. Phys.: Condens. Matter **7** 965 (1995).

[14] S. M. Reimann and M. Manninen. *Electronic structure of quantum dots*. Rev. Mod. Phys. **74** 1283 (2002).

[15] A. E. Rothman and D. A. Mazziotti. *Variational reduced-density-matrix theory applied to the electronic structure of few-electron quantum dots*. Phys. Rev. A **78** 032510 (2008).

[16] F. Hund. *Zur Deutung verwickelter Spektren, insbesondere der Elemente Scandium bis Nickel*. Z. Phys. **33** 345 (1925).

[17] F. Hund. *Zur Deutung verwickelter Spektren II*. Z. Phys. **34** 296 (1925).

[18] S. Tarucha, D. G. Austing, T. Honda, R. J. van der Hage and L. P. Kouwenhoven. *Shell Filling and Spin Effects in a Few Electron Quantum Dot*. Phys. Rev. Lett. **77** 3613 (1996).

[19] L. P. Kouwenhoven, T. H. Oosterkamp, M. W. S. Danoesastro, M. Eto, D. G. Austing, T. Honda and S. Tarucha. *Excitation Spectra of Circular, Few-Electron Quantum Dots*. Science **278** 1788 (1997).

[20] L. P. Kouwenhoven, D. G. Austing and S. Tarucha. *Few-electron quantum dots*. Rep. Prog. Phys. **64** 701 (2001).

[21] P. Matagne, J. P. Leburton, D. G. Austing and S. Tarucha. *Shell charging and spin-filling sequences in realistic vertical quantum dots*. Phys. Rev. B **65** 085325 (2002).

[22] Y. Sajeev, M. Sindelka and N. Moiseyev. *Hund's multiplicity rule: From atoms to quantum dots*. J. Chem. Phys. **128** 061101 (2008).

[23] T. Sako, J. Paldus, A. Ichimura and G. H. F. Diercksen. *Origin of the first Hund rule and the structure of Fermi holes in two-dimensional He-like atoms and two-electron quantum dots.* J. Phys. B: At. Mol. Opt. **45** 235001 (2012).

[24] J. Katriel and S. I. Themelis. *Hund's rule in the doubly excited states of the helium isoelectronic sequence.* Int. J. Quantum Chem. **112** 2880 (2012).

[25] W. Kutzelnigg and J. Morgan. *Hund's rules.* Z. Phys. D **36** 197 (1996).

[26] D. P. DiVincenzo. *Quantum Computation.* Science **270** 255 (1995).

[27] A. Barenco, C. H. Bennett, R. Cleve, D. P. DiVincenzo, N. Margolus, P. Shor, T. Sleator, J. A. Smolin and H. Weinfurter. *Elementary gates for quantum computation.* Phys. Rev. A **52** 3457 (1995).

[28] M. A. Nielsen and I. L. Chuang. *Quantum computation and quantum information* (Cambridge university press, 2010).

[29] T. Monz, P. Schindler, J. T. Barreiro, M. Chwalla, D. Nigg, W. A. Coish, M. Harlander, W. Hänsel, M. Hennrich and R. Blatt. *14-Qubit Entanglement: Creation and Coherence.* Phys. Rev. Lett. **106** 130506 (2011).

[30] Z. Bian, F. Chudak, W. G. Macready, L. Clark and F. Gaitan. *Experimental determination of Ramsey numbers with quantum annealing.* arXiv 1201.1842v2 (2012).

[31] M. Saffman, T. G. Walker and K. Mølmer. *Quantum information with Rydberg atoms.* Rev. Mod. Phys. **82** 2313 (2010).

[32] D. Loss and D. P. DiVincenzo. *Quantum computation with quantum dots.* Phys. Rev. A **57** 120 (1998).

[33] G. Burkard, D. Loss and D. P. DiVincenzo. *Coupled quantum dots as quantum gates.* Phys. Rev. B **59** 2070 (1999).

[34] A. Imamoglu, D. D. Awschalom, G. Burkard, D. P. DiVincenzo, D. Loss, M. Sherwin and A. Small. *Quantum Information Processing Using Quantum Dot Spins and Cavity QED.* Phys. Rev. Lett. **83** 4204 (1999).

[35] D. D. Awschalom, L. C. Bassett, A. S. Dzurak, E. L. Hu and J. R. Petta. *Quantum Spintronics: Engineering and Manipulating Atom-Like Spins in Semiconductors.* Science **339** 1174 (2013).

[36] M. A. Schlosshauer. *Decoherence: and the Quantum-To-Classical Transition (The Frontiers Collection)* (Springer, Berlin Heidelberg, 2010).

[37] W. H. Zurek. *Decoherence, einselection, and the quantum origins of the classical.* Rev. Mod. Phys. **75** 715 (2003).

[38] R. Horodecki, P. Horodecki, M. Horodecki and K. Horodecki. *Quantum entanglement.* Rev. Mod. Phys. **81** 865 (2009).

[39] M. C. Tichy, F. Mintert and A. Buchleitner. *Essential entanglement for atomic and molecular physics.* J. Phys. B: At. Mol. Opt. **44** 192001 (2011).

[40] J. P. Coe, A. Sudbery and I. D'Amico. *Entanglement and density-functional theory: Testing approximations on Hooke's atom.* Phys. Rev. B **77** 205122 (2008).

[41] S. Abdullah, J. P. Coe and I. D'Amico. *Effect of confinement potential geometry on entanglement in quantum dot-based nanostructures.* Phys. Rev. B **80** 235302 (2009).

[42] R. G. Nazmitdinov, N. S. Simonović, A. R. Plastino and A. V. Chizhov. *Shape transitions in excited states of two-electron quantum dots in a magnetic field.* J. Phys. B: At. Mol. Opt. **45** 205503 (2012).

[43] P. Kościk and H. Hassanabadi. *Entanglement in Hookes Law Atoms: an Effect of the Dimensionality of the Space.* Few-Body Systems **52** 189 (2012).

[44] J. S. Dehesa, T. Koga, R. J. Yáñez, A. R. Plastino and R. O. Esquivel. *Quantum entanglement in helium.* J. Phys. B: At. Mol. Opt. **45** 015504 (2012).

[45] Y.-C. Lin, C.-Y. Lin and Y. K. Ho. *Spatial entanglement in two-electron atomic systems.* Phys. Rev. A **87** 022316 (2013).

[46] D. Manzano, A. R. Plastino, J. S. Dehesa and T. Koga. *Quantum entanglement in two-electron atomic models.* J. Phys. A: Math. Gen. **43** 275301 (2010).

[47] A. P. Majtey, A. R. Plastino and J. S. Dehesa. *The relationship between entanglement, energy and level degeneracy in two-electron systems.* J. Phys. A: Math. Gen. **45** 115309 (2012).

[48] R. A. Jalabert and H. M. Pastawski. *Environment-Independent Decoherence Rate in Classically Chaotic Systems*. Phys. Rev. Lett. **86** 2490 (2001).

[49] F. M. Cucchietti, D. A. R. Dalvit, J. P. Paz and W. H. Zurek. *Decoherence and the Loschmidt Echo*. Phys. Rev. Lett. **91** 210403 (2003).

[50] B. Casabone, I. Garca-Mata and D. A. Wisniacki. *Discrepancies between decoherence and the Loschmidt echo*. Europhys. Lett. **89** 50009 (2010).

[51] P. R. Zangara, A. D. Dente, P. R. Levstein and H. M. Pastawski. *Loschmidt echo as a robust decoherence quantifier for many-body systems*. Phys. Rev. A **86** 012322 (2012).

[52] A. Peres. *Stability of quantum motion in chaotic and regular systems*. Phys. Rev. A **30** 1610 (1984).

[53] T. Gorin, T. Prosen, T. H. Seligman and M. Znidaric. *Dynamics of Loschmidt echoes and fidelity decay*. Phys. Rep. **435** 33 (2006).

[54] G. Manfredi and P.-A. Hervieux. *Loschmidt Echo in a System of Interacting Electrons*. Phys. Rev. Lett. **97** 190404 (2006).

[55] G. Manfredi and P.-A. Hervieux. *Fidelity Decay in Trapped Bose-Einstein Condensates*. Phys. Rev. Lett. **100** 050405 (2008).

[56] G. Manfredi and P.-A. Hervieux. *Loschmidt echo for the many-electron dynamics in nonparabolic quantum wells*. New J. Phys. **11** 013050 (2009).

[57] O. Bohigas, M. J. Giannoni and C. Schmit. *Characterization of Chaotic Quantum Spectra and Universality of Level Fluctuation Laws*. Phys. Rev. Lett. **52** 1 (1984).

[58] M. C. Gutzwiller. *From classical to quantum mechanics with hard chaos*. J. Phys. Chem. **92** 3154 (1988).

[59] G. Tanner, K. Richter and J.-M. Rost. *The theory of two-electron atoms: between ground state and complete fragmentation*. Rev. Mod. Phys. **72** 497 (2000).

[60] T. A. Brody, J. Flores, J. B. French, P. A. Mello, A. Pandey and S. S. M. Wong. *Random-matrix physics: spectrum and strength fluctuations*. Rev. Mod. Phys. **53** 385 (1981).

[61] M. V. Berry and M. Robnik. *Semiclassical level spacings when regular and chaotic orbits coexist.* J. Phys. A: Math. Gen. **17** 2413 (1984).

[62] T. Prosen. *Berry-Robnik level statistics in a smooth billiard system.* J. Phys. A: Math. Gen. **31** 7023 (1998).

[63] V. A. Podolskiy and E. E. Narimanov. *Level spacing distribution in systems with partially chaotic classical dynamics.* Phys. Lett. A **362** 412 (2007).

[64] A. Bäcker, R. Ketzmerick, S. Löck and N. Mertig. *Fractional-Power-Law Level Statistics Due to Dynamical Tunneling.* Phys. Rev. Lett. **106** 024101 (2011).

[65] T. Rudolf, N. Mertig, S. Löck and A. Bäcker. *Consequences of flooding on spectral statistics.* Phys. Rev. E **85** 036213 (2012).

[66] J. Eiglsperger, T. Kramer and J. Madroñero. *Fluctuations in the spectra of open few-body systems.* New J. Phys. **13** 063033 (2011).

[67] J. Laskar. *The chaotic motion of the solar system: A numerical estimate of the size of the chaotic zones.* Icarus **88** 266 (1990).

[68] E. N. Lorenz. *Deterministic Nonperiodic Flow.* J. Atmos. Sci. **20** 130 (1963).

[69] R. L. Kautz and R. Monaco. *Survey of chaos in the rf-biased Josephson junction.* J. Appl. Phys. **57** 875 (1985).

[70] M. Rigol, V. Dunjko and M. Olshanii. *Thermalization and its mechanism for generic isolated quantum systems.* Nature **452** 854 (2008).

[71] L. F. Santos, F. Borgonovi and F. M. Izrailev. *Chaos and Statistical Relaxation in Quantum Systems of Interacting Particles.* Phys. Rev. Lett. **108** 094102 (2012).

[72] L. F. Santos, A. Polkovnikov and M. Rigol. *Weak and strong typicality in quantum systems.* Phys. Rev. E **86** 010102 (2012).

[73] M. Rigol and M. Srednicki. *Alternatives to Eigenstate Thermalization.* Phys. Rev. Lett. **108** 110601 (2012).

[74] Y. G. Sinai. *Dynamical systems with elastic reflections.* Russ. Math. Surv. **25** 137 (1970).

[75] M. Robnik. *Classical dynamics of a family of billiards with analytic boundaries.* J. Phys. A: Math. Gen. **16** 3971 (1983).

[76] M. Robnik. *Quantising a generic family of billiards with analytic boundaries.* J. Phys. A: Math. Gen. **17** 1049 (1984).

[77] L. A. Bunimovich and L. V. Vela-Arevalo. *Many faces of stickiness in Hamiltonian systems.* Chaos **22** 026103 (2012).

[78] H.-J. Stöckmann and J. Stein. *"Quantum" chaos in billiards studied by microwave absorption.* Phys. Rev. Lett. **64** 2215 (1990).

[79] H.-J. Stöckmann. *Quantum Chaos: An Introduction* (Cambridge University Press, Cambridge, 2007).

[80] L. Toporowicz and M. Beims. *Correlation effects of two interacting particles in a circular billiard.* Physica A **371** 5 (2006).

[81] S. Sawada, A. Terai and K. Nakamura. *Coulomb-interaction-induced quantum irregularity in two electrons within a hard-walled circular billiard.* Chaos Solitons Fractals **40** 862 (2009).

[82] P. L. J. Solanpää, J. Nokelainen and E. Räsänen. *Chaotic properties of Coulomb-interacting circular billiards.* arXiv 1210.2811v2 (2013).

[83] K. Nakamura and T. Harayama. *Quantum Chaos and Quantum Dots (Mesoscopic Physics and Nanotechnology)* (Oxford University Press, USA, 2004).

[84] D. K. Ferry, A. M. Burke, R. Akis, R. Brunner, T. E. Day, R. Meisels, F. Kuchar, J. P. Bird and B. R. Bennett. *Open quantum dots – probing the quantum to classical transition.* Semicond. Sci. Technol **26** 043001 (2011).

[85] E. Altmann, J. S. E. Portela and T. Tél. *Leaking chaotic systems.* arXiv 1208.0254v2 (2013).

[86] M. Wagner, U. Merkt and A. V. Chaplik. *Spin-singlet – spin-triplet oscillations in quantum dots.* Phys. Rev. B **45** 1951 (1992).

[87] D. Pfannkuche and R. R. Gerhardts. *Quantum-dot helium: Effects of deviations from a parabolic confinement potential.* Phys. Rev. B **44** 13132 (1991).

[88] S. Bell, R. Davidson and P. A. Warsop. *The one- and two-dimensional quartic oscillators.* J. Phys. B: At. Mol. Opt. **3** 113 (1970).

[89] A. I. Ekimov and A. A. Onushchenko. *Size quantization of the electron energy spectrum in a microscopic demiconductor crystal.* J. Exp. Theor. Phys. Lett. **40** 337 (1984).

[90] R. Rossetti, R. Hull, J. M. Gibson and L. E. Brus. *Excited electronic states and optical spectra of ZnS and CdS crystallites in the $\approx$ 15 to 50 Å size range: Evolution from molecular to bulk semiconducting properties.* J. Chem. Phys. **82** 552 (1985).

[91] C. Sikorski and U. Merkt. *Spectroscopy of electronic states in InSb quantum dots.* Phys. Rev. Lett. **62** 2164 (1989).

[92] R. C. Ashoori, H. L. Stormer, J. S. Weiner, L. N. Pfeiffer, S. J. Pearton, K. W. Baldwin and K. W. West. *Single-electron capacitance spectroscopy of discrete quantum levels.* Phys. Rev. Lett. **68** 3088 (1992).

[93] B. Meurer, D. Heitmann and K. Ploog. *Single-electron charging of quantum-dot atoms.* Phys. Rev. Lett. **68** 1371 (1992).

[94] T. Demel, D. Heitmann, P. Grambow and K. Ploog. *Nonlocal dynamic response and level crossings in quantum-dot structures.* Phys. Rev. Lett. **64** 788 (1990).

[95] W. Kohn. *Cyclotron Resonance and de Haas-van Alphen Oscillations of an Interacting Electron Gas.* Phys. Rev. **123** 1242 (1961).

[96] W. G. van der Wiel, T. H. Oosterkamp, J. W. Janssen, L. P. Kouwenhoven, D. G. Austing, T. Honda and S. Tarucha. *Singlet-triplet transitions in a few-electron quantum dot.* Physica B **256-258** 173 (1998).

[97] S. Tarucha, T. Honda, D. G. Austing, Y. Tokura, K. Muraki, T. H. Oosterkamp, J. W. Janssen and L. P. Kouwenhoven. *Electronic states in quantum dot atoms and molecules.* Physica E **3** 112 (1998).

[98] R. C. Ashoori, H. L. Stormer, J. S. Weiner, L. N. Pfeiffer, K. W. Baldwin and K. W. West. *N-electron ground state energies of a quantum dot in magnetic field.* Phys. Rev. Lett. **71** 613 (1993).

[99] T. Schmidt, M. Tewordt, R. H. Blick, R. J. Haug, D. Pfannkuche, K. v. Klitzing, A. Förster and H. Lüth. *Quantum-dot ground states in a magnetic*

field studied by single-electron tunneling spectroscopy on double-barrier heterostructures. Phys. Rev. B **51** 5570 (1995).

[100] G. Thurner, H. Herold, H. Ruder, G. Schlicht and G. Wunner. *Note on binding energies of helium-like systems in magnetic fields*. Phys. Lett. A **89** 133 (1982).

[101] R. G. Nazmitdinov, N. S. Simonović and J. M. Rost. *Semiclassical analysis of a two-electron quantum dot in a magnetic field: Dimensional phenomena*. Phys. Rev. B **65** 155307 (2002).

[102] P. Matagne and J.-P. Leburton. *Three-dimensional analysis of the electronic structure of cylindrical vertical quantum dots*. Phys. Rev. B **65** 235323 (2002).

[103] N. R. Kestner and O. Sinanoḡlu. *Study of Electron Correlation in Helium-Like Systems Using an Exactly Soluble Model*. Phys. Rev. **128** 2687 (1962).

[104] D. P. O'Neill and P. M. W. Gill. *Wave functions and two-electron probability distributions of the Hooke's-law atom and helium*. Phys. Rev. A **68** 022505 (2003).

[105] S. Kais, D. R. Herschbach, N. C. Handy, C. W. Murray and G. J. Laming. *Density functionals and dimensional renormalization for an exactly solvable model*. J. Chem. Phys. **99** 417 (1993).

[106] T. Sako, J. Paldus and G. H. F. Diercksen. *Origin of Hund's multiplicity rule in quasi-two-dimensional two-electron quantum dots*. Phys. Rev. A **81** 022501 (2010).

[107] V. Fock. *Bemerkung zur Quantelung des harmonischen Oszillators im Magnetfeld*. Z. Phys. **47** 446 (1928).

[108] M. Taut. *Two electrons in an external oscillator potential: Particular analytic solutions of a Coulomb correlation problem*. Phys. Rev. A **48** 3561 (1993).

[109] M. Taut. *Two electrons in a homogeneous magnetic field: particular analytical solutions*. J. Phys. A: Math. Gen. **27** 1045 (1994).

[110] A. Turbiner. *Two electrons in an external oscillator potential: The hidden algebraic structure*. Phys. Rev. A **50** 5335 (1994).

[111] B. S. Kandemir. *Two interacting electrons in a uniform magnetic field and a parabolic potential: The general closed-form solution*. J. Math. Phys. **46** 032110 (2005).

[112] S. Klama and E. G. Mishchenko. *Two electrons in a quantum dot: a semiclassical approach*. J. Phys.: Condens. Matter **10** 3411 (1998).

[113] T. Kramer. *Two interacting electrons in a magnetic field: comparison of semiclassical, quantum, and variational solutions*. AIP Conf. Proc. **1323** 178 (2010).

[114] T. Sako, P.-A. Hervieux and G. H. F. Diercksen. *Distribution of oscillator strength in Gaussian quantum dots: An energy flow from center-of-mass mode to internal modes*. Phys. Rev. B **74** 045329 (2006).

[115] J. Eiglsperger. *Highly doubly excited states of two-electron atoms: spectra, cross sections and localization properties*. Ph.D. thesis, Technische Universität München (2010). http://nbn-resolving.de/urn/resolver.pl?urn:nbn:de:bvb:91-diss-20100607-972994-1-9.

[116] P. A. Maksym and T. Chakraborty. *Quantum dots in a magnetic field: Role of electron-electron interactions*. Phys. Rev. Lett. **65** 108 (1990).

[117] O. Ciftja and M. G. Faruk. *Two-dimensional quantum-dot helium in a magnetic field: Variational theory*. Phys. Rev. B **72** 205334 (2005).

[118] J. Madroñero. *Spectral properties of planar helium under periodic driving*. Ph.D. thesis, Ludwig-Maximilians-Universität München (2004). http://nbn-resolving.de/urn:nbn:de:bvb:19-21879.

[119] J. Eiglsperger and J. Madronero. *Highly doubly excited states of planar helium: Fluctuations in photoionization cross sections*. Phys. Rev. A **80** 022512 (2009).

[120] J. Madroñero and B. Piraux. *Explicit time-propagation method to treat the dynamics of driven complex systems*. Phys. Rev. A **80** 033409 (2009).

[121] C. Lanczos. *An iteration method for the solution of the eigenvalue problem of linear differential and integral operators*. J. Res. Natl. Bur. Stand. **45** 255 (1950).

[122] T. Ericsson and A. Ruhe. *The Spectral Transformation Lanczos Method for the Numerical Solution of Large Sparse Generalized Symmetric Eigenvalue Problems*. Math. Comp. **35** 1251 (1980).

[123] J. M. Sanz-Serna and M. Calvo. *Numerical Hamiltonian Problems (Applied Mathematics)* (Chapman and Hall/CRC, London, 1994).

[124] J. Laskar. *Frequency analysis for multi-dimensional systems. Global dynamics and diffusion*. Physica D **67** 257 (1993).

[125] B. Cordani. *Frequency modulation indicator, Arnolds web and diffusion in the Stark-Quadratic-Zeeman problem*. Physica D **237** 2797 (2008).

[126] J. Laskar, C. Froeschl and A. Celletti. *The measure of chaos by the numerical analysis of the fundamental frequencies. Application to the standard mapping*. Physica D **56** 253 (1992).

[127] M. Schönwetter. *A study of the complex dynamics in few-dimensional classical systems: two-electron quantum dots and coupled standard maps*. Diploma thesis, Technische Universität München (2011).

[128] R. M. May. *Simple mathematical models with very complicated dynamics*. Nature **261** 459 (1976).

[129] B. V. Chirikov. *A universal instability of many-dimensional oscillator systems*. Phys. Rep. **52** 263 (1979).

[130] P. Cvitanović, R. Artuso, R. Mainieri, G. Tanner and G. Vattay. *Chaos: Classical and Quantum* (Niels Bohr Institute, Copenhagen, 2012).

[131] H. Goldstein, C. P. Poole, Jr. and J. L. Safko, Sr. *Klassische Mechanik (German Edition)* (Wiley-VCH, Weinheim, 2006).

[132] A. Lichtenberg and M. Lieberman. *Regular and chaotic dynamics* (Springer, Berlin Heidelberg, 1992).

[133] L. Chierchia and J. N. Mather. *Kolmogorov-Arnold-Moser theory*. Scholarpedia **5** 2123 (2010).

[134] M. C. Gutzwiller. *Phase-Integral Approximation in Momentum Space and the Bound States of an Atom*. J. Math. Phys. **8** 1979 (1967).

[135] H. Weyl. *Ueber die asymptotische Verteilung der Eigenwerte.* D. Hilbert (editor), *Nachrichten von der Gesellschaft der Wissenschaften zu Göttingen, Mathematisch-Physikalische Klasse*, volume 2 of *Lecture Notes in Physics*, 110 (Vandenhoeck & Ruprecht, Göttingen, 1911).

[136] O. Bohigas and M.-J. Giannoni. *Chaotic motion and random matrix theories.* J. Dehesa, J. Gomez and A. Polls (editors), *Mathematical and Computational Methods in Nuclear Physics*, volume 209 of *Lecture Notes in Physics*, 1 – 99 (Springer, Berlin Heidelberg, 1984).

[137] F. Haake. *Quantum Signatures of Chaos* (Springer, Berlin Heidelberg, 2010).

[138] L. Mehta. *Random Matrices* (Academic Press, 1991).

[139] V. Zelevinsky, B. Brown, N. Frazier and M. Horoi. *The nuclear shell model as a testing ground for many-body quantum chaos.* Phys. Rep. **276** 85 (1996).

[140] E. P. Wigner. *Characteristic Vectors of Bordered Matrices With Infinite Dimensions.* Ann. of Math. **62** 548 (1955).

[141] M. J. Davis and E. J. Heller. *Quantum dynamical tunneling in bound states.* J. Chem. Phys. **75** 246 (1981).

[142] S. Tomsovic and D. Ullmo. *Chaos-assisted tunneling.* Phys. Rev. E **50** 145 (1994).

[143] V. A. Podolskiy and E. E. Narimanov. *Semiclassical Description of Chaos-Assisted Tunneling.* Phys. Rev. Lett. **91** 263601 (2003).

[144] X. Yang and J. Burgdörfer. *Statistics of avoided crossings for generic quantum systems.* Phys. Rev. A **48** 83 (1993).

[145] P. Plötz, M. Lubasch and S. Wimberger. *Detection of avoided crossings by fidelity.* Physica A **390** 1363 (2011).

[146] P. Plötz. *Complex Dynamics of Ultracold Atoms.* Ph.D. thesis, Ruprecht-Karls-Universität Heidelberg (2010). http://www.ub.uni-heidelberg.de/archiv/11123.

[147] H. Venzl, T. Zech, B. Ole, M. Hiller, F. Mintert and A. Buchleitner. *Solitonic eigenstates of the chaotic BoseHubbard Hamiltonian.* Appl. Phys. B **98** 647 (2010).

[148] J. Emerson, Y. S. Weinstein, S. Lloyd and D. G. Cory. *Fidelity Decay as an Efficient Indicator of Quantum Chaos*. Phys. Rev. Lett. **89** 284102 (2002).

[149] R. Jozsa and B. Schumacher. *A New Proof of the Quantum Noiseless Coding Theorem*. J. Mod. Optic. **41** 2343 (1994).

[150] R. Jozsa. *Fidelity for Mixed Quantum States*. J. Mod. Optic. **41** 2315 (1994).

[151] P. Jacquod and C. Petitjean. *Decoherence, entanglement and irreversibility in quantum dynamical systems with few degrees of freedom*. Adv. Phys. **58** 67 (2009).

[152] T. Prosen and M. Znidaric. *Quantum freeze of fidelity decay for a class of integrable dynamics*. New J. Phys. **5** 109 (2003).

[153] T. c. v. Prosen and M. Žnidarič. *Quantum Freeze of Fidelity Decay for Chaotic Dynamics*. Phys. Rev. Lett. **94** 044101 (2005).

[154] Private communications with P.-A. Hervieux and G. Manfredi, Institut de Physique et Chimie des Matériaux, CNRS and Université de Strasbourg, Strasbourg (2010).

[155] D. Cohen and E. J. Heller. *Unification of Perturbation Theory, Random Matrix Theory, and Semiclassical Considerations in the Study of Parametrically Dependent Eigenstates*. Phys. Rev. Lett. **84** 2841 (2000).

[156] S. O. Fatunla. *An Implicit Two-Point Numerical Integration Formula for Linear and Nonlinear Stiff Systems of Ordinary Differential Equations*. Math. Comput. **32** 1 (1978).

[157] S. O. Fatunla. *Numerical Integrators for Stiff and Highly Oscillatory Differential Equations*. Math. Comput. **34** 373 (1980).

[158] W. E. Arnoldi. *The principle of minimized iteration in the solution of the matrix eigenvalue problem*. Quart. Appl. Math. **9** 17 (1951).

[159] E. S. Smyth, J. S. Parker and K. Taylor. *Numerical integration of the time-dependent Schrdinger equation for laser-driven helium*. Comput. Phys. Commun. **114** 1 (1998).

[160] C. Liu and M. Nisoli. *Complete characterization of a coherent superposition of atomic states by asymmetric attosecond photoionization.* Phys. Rev. A **85** 053423 (2012).

[161] R. Pazourek, J. Feist, S. Nagele and J. Burgdörfer. *Attosecond Streaking of Correlated Two-Electron Transitions in Helium.* Phys. Rev. Lett. **108** 163001 (2012).

[162] A. Hamido, J. Eiglsperger, J. Madroñero, F. Mota-Furtado, P. O'Mahony, A. L. Frapiccini and B. Piraux. *Time scaling with efficient time-propagation techniques for atoms and molecules in pulsed radiation fields.* Phys. Rev. A **84** 013422 (2011).

[163] P. J. van der Houwen; B. P. Sommeijer; W. Couzy. *Embedded diagonally implicit Runge-Kutta algorithms on parallel computers.* Math. Comput. 135 – 159 (1992).

[164] S. Perez-Rodriguez, S. Gonzalez-Pinto and B. Sommeijer. *An iterated Radau method for time-dependent PDEs.* J. Comp. Appl. Math. **231** 49 (2009).

[165] A. Meister. *Numerik linearer Gleichungssysteme* (Vieweg+Teubner, Wiesbaden, 2007).

[166] Private communications with B. Piraux, Université catholique de Louvain, Louvain–La–Neuve (2012).

[167] Private communications with F. Mota–Furtado and P. O'Mahony, Royal Holloway University of London, London (2012).

[168] J. D. Morgan and W. Kutzelnigg. *Hund's rules, the alternating rule, and symmetry holes.* J. Phys. Chem. **97** 2425 (1993).

[169] J. C. Slater. *The Theory of Complex Spectra.* Phys. Rev. **34** 1293 (1929).

[170] W. Heisenberg. *Mehrkörperproblem und Resonanz in der Quantenmechanik.* Z. Phys. **38** 411 (1926).

[171] P. A. M. Dirac. *On the Theory of Quantum Mechanics.* Proc. R. Soc. Lond. A **112** 661 (1926).

[172] E. R. Davidson. *Single-Configuration Calculations on Excited States of Helium. II.* J. Chem. Phys. **42** 4199 (1965).

[173] Y. Accad, C. L. Pekeris and B. Schiff. *S and P States of the Helium Isoelectronic Sequence up to Z = 10*. Phys. Rev. A **4** 516 (1971).

[174] E. A. Colbourn and C. A. Coulson. *A 1/Z-expansion analysis of singlet-triplet differences in the lowest $^{1,3}\Pi$ states of H_2 , HeH^+, and LiH^{++}*. J. Phys. B: At. Mol. Opt. **7** 1574 (1974).

[175] R. J. Boyd and C. A. Coulson. *Coulomb hole in some excited states of helium*. J. Phys. B: At. Mol. Opt. **6** 782 (1973).

[176] D. A. Kohl. *On the Difference in Energy between Singlet and Triplet States*. J. Chem. Phys. **56** 4236 (1972).

[177] H. Tatewaki and K. Tanaka. *Correlation holes and wavefunctions for the ^1S, $^1P^o$, and $^3P^o$ states of Be, B^+, and C^{2+}*. J. Chem. Phys. **60** 601 (1974).

[178] J. Colpa and M. Islip. *Hund's rule and the Z expansion for the energy, electron repulsion and electron nuclear attraction*. Mol. Phys. **25** 701 (1973).

[179] J. Colpa, A. J. Thakkar, V. H. Smith and P. Randle. *An analysis of energy differences in atomic multiplets in connection with the inequality formulation of Hund's rules*. Mol. Phys. **29** 1861 (1975).

[180] R. J. Boyd. *A quantum mechanical explanation for Hund's multiplicity rule*. Nature **310** 480 (1984).

[181] H. N. Russell and W. F. Meggers. *An analysis of the arc and spark spectra of scandium (Sc I and Sc II)*. Sci. Pap. Bur. Stand. **22** 329 (1927).

[182] L. Amico, R. Fazio, A. Osterloh and V. Vedral. *Entanglement in many-body systems*. Rev. Mod. Phys. **80** 517 (2008).

[183] G. Ghirardi, L. Marinatto and T. Weber. *Entanglement and Properties of Composite Quantum Systems: A Conceptual and Mathematical Analysis*. J. Stat. Phys. **108** 49 (2002).

[184] G. Ghirardi and L. Marinatto. *General criterion for the entanglement of two indistinguishable particles*. Phys. Rev. A **70** 012109 (2004).

[185] A. R. Plastino, D. Manzano and J. S. Dehesa. *Separability criteria and entanglement measures for pure states of N identical fermions*. Europhys. Lett. **86** 20005 (2009).

[186] L. Wendler, V. M. Fomin, A. V. Chaplik and A. O. Govorov. *Optical properties of two interacting electrons in quantum rings: Optical absorption and inelastic light scattering.* Phys. Rev. B **54** 4794 (1996).

[187] E. Huens. *Interaction de l'hydrogène atomique avec une impulsion laser courte et intense: développement d'une méthode numérique d'analyse et application àl'étude de la stabilisation.* Ph.D. thesis, Université catholique de Louvain (1996).

I want morebooks!

Buy your books fast and straightforward online - at one of the world's fastest growing online book stores! Environmentally sound due to Print-on-Demand technologies.

Buy your books online at
www.get-morebooks.com

Kaufen Sie Ihre Bücher schnell und unkompliziert online – auf einer der am schnellsten wachsenden Buchhandelsplattformen weltweit! Dank Print-On-Demand umwelt- und ressourcenschonend produziert.

Bücher schneller online kaufen
www.morebooks.de

OmniScriptum Marketing DEU GmbH
Heinrich-Böcking-Str. 6-8
D - 66121 Saarbrücken

Telefax: +49 681 93 81 567-9

info@omniscriptum.de
www.omniscriptum.com

Printed by Books on Demand GmbH, Norderstedt / Germany